DREAM

THE ART AND SCIENCE OF SLUMBER

SCOTT CARNEY

Published by Foxtopus, Inc.

Copyright ©2024 by Scott Carney

Foxtopus supports copyright. Copyright fuels creativity, encourages diverse voices, promotes free speech, and creates a vibrant culture. Thank you for buying an authorized edition of this book, and for not reproducing, scanning, or distributing any part of it without permission. You are supporting writers and allowing Foxtopus to continue to publish books for every reader.

Book design by Scott Carney

LIBRARY OF CONGRESS CATALOGING-IN-PUBLICATION DATA
Carney, Scott, 1978– author.

ISBN

Also by Scott Carney

The Red Market: *On the Trail of the World's Organ Brokers, Bone Thieves, Blood Farmers, and Child Traffickers*

The Enlightenment Trap: *Obsession, Madness, and Death on Diamond Mountain*

What Doesn't Kill Us: *How Freezing Water, Extreme Altitude and Environmental Conditioning Will Renew Our Lost Evolutionary Strength*

The Wedge: *Evolution, Consciousness, Stress, and the Key to Human Resilience*

The Vortex: *A Trues Story of History's Deadliest Storm, an Unspeakable War, and Liberation* (with Jason Miklian)

For the dreamers

1

INTRODUCTION

When God created the first human, he told him what to expect from life. He said the man would live about eighty years but spend about a third of that time asleep in bed. The man looked back at God, annoyed by the futility of so much wasted time. But God wasn't finished, and he leaned in to whisper in the man's ear: "And that's going to be your favorite part."

At first glance, neither sleeping nor dreaming makes any sense. What is the point of closing your eyes, letting down all of your defenses and stopping any activity that might further your chances of survival—just so you can bask in disconnected images of nonsense that you are almost definitely not going to remember for more than a few minutes after you wake up? Why does every single living creature, plant and insect go through cyclical periods of non-activity? Scientists, biologists, doctors, anthropologists, theologians and philosophers have chewed on these questions for millennia. While they have each posited their own varying opinions over the ages, the only thing that is certain is that whatever happens during sleep is absolutely necessary. Even dreaming is vital.

According to one strain of clinical thinking—as well as guidelines laid down by the Centers for Disease Control—people should aim to get seven to nine hours of uninterrupted sleep a day. Consistently failing to achieve that mark initiates a cascade of well-understood health problems that begin with brain fog, chronic anxiety, and weight gain and lead to, among other things, a compromised immune system, cancer, and even death. Sleep benefits the brain and the body.

Sleep is vital to a healthy waking life, and the different stages of sleep—REM (rapid eye movement), NREM (non-rapid eye movement), light and deep sleep—all play key roles in consciousness, motor learning and recovery. For more than half a century, sleep researchers have conducted experiments on nighttime sleep patterns, carefully dissecting the various stages of sleep and correlating those stages with cognitive and health outcomes. Only a handful of laboratory studies seriously consider the content and experience of dreaming. Researchers can study the physiology of a person as they move between different dream states, but they don't have many useful tools to understand the subjective experience of dreaming. For the most part, we're terrible at remembering dreams.

This is at least somewhat surprising, because some of the greatest minds in history credit the innovative ways that they moved between sleep and wakefulness as the secret to their success. Benjamin Franklin and Winston Churchill made access to dreams (through innovative napping protocols) an essential part of their daily routine—not only because it made them feel better, but because it helped them come up with new and useful ideas. Napoleon napped before every battle he fought except for Waterloo (which he lost). Thomas Jefferson and Salvador Dalí directly tapped into their dreams with a simple technique to wake themselves up in the middle of naps in order to harness the implicit creative thinking that occurs in the dream space.

The revered Stanford sleep scientist William Dement was a rare champion of napping, and was so diligent about his practice that he would occasionally leave class mid-lecture or drop out of tenure meetings to keep his strict restorative schedule. In retrospect, this might have been one of the reasons that Dement remained mentally sharp until the end of his days at the age of 91. For these luminaries, profound things happened as they transitioned between consciousness and unconsciousness.

In the past hundred years or so, there have been many books on dreaming. Most notably, Sigmund Freud and Carl Jung looked to dreams to inform their analysis of hidden mental processes and, in Jung's case, even a collective human unconscious. More recent researchers discarded most of this pioneering work in favor of laboratory measurements of physiological changes that happen during different sleeping states, mostly discarding the content of dreams as meaningless nonsense. Meanwhile, other writers—usually not scientists—have taken an alternate tack and examined the meaning of dream archetypes writ large. They can tell you when a cigar is definitely not a cigar, and what the appearance of your childhood home in a dream means in terms of lingering issues with your parents. Still other books examine lucid dreaming, where a sleeper wakes up in a dream and is able to control the physics of the world around them.

All of these additions to the sleep literature are interesting, and even potentially useful, but they miss something that I think is even more important about dreams, something so obvious that we've simply lost the thread: that who we are in our dreams is actually an extension of who we are when we're awake. We are accustomed to thinking of these two mental and physical spaces as somehow entirely separate from one another, like polar opposites that never touch. As we shall see, however, they really aren't that different at all. Instead of an on-off switch, it's more useful to think of these states as polarities that continuously flow between one another. In many ways, what we think of as "sleeping" correlates with more brain activity (and thus likely experiences) than what we feel while we are "awake." The distinction between what we think of as consciousness and unconsciousness is going to get very blurry.

Prior to industrialization, Western countries had widely diverse sleeping patterns that included not only siestas, but also two periods of sleep in the middle of the night, accompanied by a few hours of wakefulness during the darkest hours. That Western tradition continues inside a few ancient religious orders that still retain the practice of "second sleep," which they note is often an ideal time for prayer and spiritual insight. This entire mode of consciousness exists in the space where dreaming and waking overlap. Outside of any religious implications, anyone who has suffered from a bout of insomnia knows how different nighttime thinking is from the focused daytime mind. While most of us prefer to avoid insomnia, who hasn't noticed how their thoughts after midnight allow them to look at daytime problems in different ways? During that witching hour, I often come up with action plans and solutions that aren't always obvious. This is because, even though you lie awake at night, circadian rhythms in the brain blend the benefits of dreaming into your semi-conscious thoughts.

Both historical and ethnographic data sets show that modern-day sleep patterns are not only an aberration from evolutionary norms, but also profoundly linked to chronic exhaustion and systematic exploitation.

Scholars Bell Hooks and activists like Tricia Hersey of The Nap Ministry trace the roots of our modern disregard for natural sleep patterns to the brutal conditions on slave plantations, where colonial (and white) masters invented a new economic system that measured human productivity with machine-like precision. Those plantations devalued human lives in favor of raw metrics of industrial output. They argue that the racial disparities of today, at least in part, lie in the historical sleep deprivation of black ancestors. While the worst period of chattel slavery is behind us, the economic system that we use today was, nonetheless, built on its foundations and implicitly pushes forward an agenda of dehumanization. The methods of the modern workplace coerce us into placing productivity above our own best interests. We begin to learn the importance of productivity from the moment we enter into the educational system and get our first homework assignment. The indoctrination continues relentlessly, all the way to the time that some of us are fortunate enough to retire. It's only in retirement that society agrees that we have permission to catch up on our "well-deserved rest." (The poor, for their part, may never earn the benefits of rest in retirement.)

Unfortunately, even retirement isn't quite what it's cracked up to be. After a hard-charging life of non-stop accomplishment, many retirees have simply no idea what to do with all of that empty time. For our entire adult lives, we attempt to accumulate enough wealth to finally afford rest, only to discover that we never learned the ability to relax and dream in the first place. When our nervous systems spend decades learning only one way to exist in the world, they lack the plasticity they need to slow down when we need it. Some of us even come to believe that the output of our labor defines the entire value of our lives.

While almost every one of us understands what it feels like to be tired, and that getting sleep is the only real antidote to the problem, very few of us actually think deeply about what happens during those most peaceful times of the day. While we might remember snippets of dreams in the first few minutes after waking up, we don't generally think that the very process of dreaming might be necessary to the way that we function the rest of the day. And if we do sometimes consider the power of a particular dream, we mostly let the insight go by the time we get to breakfast. Our society has a collective inability to rest and dream. Scientists often explain away the mysterious and non-narrative contents of dreams as random electrical firings in the brain that lack any coherent meaning.

Even the most ardent sleep supporters in the neuroscience community can't escape the language of economics when talking about exhaustion. According to scientific nomenclature, every human requires a certain amount of sleep every night that varies only slightly by genetics. For every hour of sleep you miss, a would-be sleeper racks up an hour of "sleep debt." If you miss an hour every night for a week, you'll be on the hook for seven hours of payback. Much like our would-be retirees with overtaxed credit cards who teeter on financial ruin, the language around sleep debt frames exhaustion and the accompanying chronic illness as a sort of bankruptcy.

While truly exhausted people have to sleep, applying economic logic to sleep patterns is, at the very least, unimaginative. Observe just about any animal in its native environment and you will see three interlocking sleep and alertness cycles. The first is the circadian rhythm that tracks along with the 24 hours of the day. Sunshine initiates our bodies to produce a careful cocktail of hormones and neurotransmitters that makes us more alert and then slowly declines throughout the day. By nighttime, a different biochemical process makes us feel sleepy as our bodies wind down to rest and repair.

The second cycle tracks along with the rotation and tilt of the planet around the sun as seasonal changes make the days longer and warmer or shorter and colder. Modern humans in affluent countries almost completely ignore seasonal cycles of light and darkness in favor of the eternal summer offered by electric lighting and indoor climate control. It takes an almost unthinkable, herculean effort to imagine conforming your sleep schedule away from the demands of modern society and instead embracing natural periods of darkness. Doing so is such a radical notion that I don't personally know a single person who is able to maintain such a schedule for more than a few days, despite the fact that it was the norm for our ancestors more than a few hundred years ago and backward to eternity.

The third cycle is the natural ebbing and flowing of alertness during daylight hours. Attention moves in and out of focus like a child playing with the dials on a telescope. While people progress through different states of alertness throughout the day, depending on their genetics, culture, diet, latitude and a thousand other factors, almost no one stays in the same steady state of focus from the moment they wake up to the moment they close their eyes at night. Instead, we wake up, consume a meal or coffee at some point, focus at work, daydream, consider our flagging attention span, switch our focus to new topics, and on and on and on. The quality of our attention varies drastically, depending on the specific thing we are trying to engage with. The blend of different conscious states of focus is, perhaps unintuitively for the Western mind, all a continuation of dream time.

We are accustomed to thinking that dreams happen when we are asleep, and we focus when we're awake, but even a cursory examination of our mental activity throughout the day proves how uselessly binary that thought really is. A more realistic description would plot wakefulness and dreaming on a continuum between focus and abstraction—or, if you prefer, between directed effort to accomplish real-world objectives and making free associations between seemingly unrelated ideas.

In a similar economical logic that puts the exhausted into sleep debt, individuals who don't conform to the focused side of the alertness continuum often end up with a diagnosis of Attention Deficit Disorder (ADD)—being treated like delinquent nations unable to manage their own economies. Yet as Malcolm Gladwell notes in his book *Outliers*, ADD diagnoses rank disproportionately high on both extremes of the social spectrum. CEOs, entrepreneurs, and creative professionals often leverage abstract thought to financial gain and professional success if they cultivate their neurodivergence in a productive direction. At the same time, ADD also plays a statistically significant role in individuals filing for bankruptcy and dropping out of high school. Additionally, while only 5 to 7% of people qualify for an ADD diagnosis, a whopping one in four American prisoners have the condition. Although there is only limited research on the dream life of people with ADD, my 45 years living with it suggest that both the advantages and disadvantages of ADD depend on societal expectations. The same is true for dreaming and our relationship with sleep in general.

Many scientific experiments and anthropological observations note that almost all of us flag approximately eight hours after we wake up in the morning. That's when our spirits begin to dwindle, our focus blurs, and problems we could have solved easily in the morning turn more perplexing. The exact timing varies slightly, according to the individual, but it tends to correspond with a dip in melatonin and an increase of the fatigue-feeling hormone adenosine. In the Western world, when two or three o'clock rolls around, we check our watches and sigh that the work day is not nearly over yet. If we're conditioned by the insanity of our relentless society, we urge our bodies forward and push the distracting sensations of fatigue away with grit and determination in order to get on with the day. This process works for a little while. We all have the ability to override the body's short-term sleep signaling. But the effort comes with a price.

The counterintuitive thing about our tendency to ignore our body's internal signaling and habitually push through to meet the demands of the world is that, if we actually paid attention to those sensations instead of our immediate objectives, we would probably do a whole lot better at actually accomplishing the very goals that stymie us.

I've written about this concept before in several other books and describe the ability to insert our conscious minds and intentions between the signals and sensations we feel from our nervous systems as a sort of "wedge." This powerful aspect of consciousness is what you could call a literal manifestation of mind over matter. In those books, I've shown how we can use Wedge techniques to increase resilience in the face of a variety of external stresses—

allowing our bodies to thrive in extreme cold and heat as well as keeping calm in the presence of dangerous conditions. But *The Wedge* also has a dark aspect. The thought to resist the sensations of fatigue corresponds with a subtle initiation of the fight-or-flight system (otherwise known as the sympathetic nervous system) and pairs with the release of energy-increasing and pain-decreasing hormones like adrenaline and cortisol. Done infrequently, it's not much of a problem. But ignore your body's commands on a regular basis and you will reap consequences in the form of constantly overactive anxiety and autoimmune conditions that come with the territory of hypervigilance. The ability to control yourself in this way is what I think of as a "negative wedge"—akin to the mental control that an anorexic extends over their sensations to hunger to stop eating the food their body needs.

It shouldn't be that surprising that we actually start feeling better about most everything when we listen to the wisdom of our bodies. This wisdom comes to us in the form of sensation. Every sensation offers us a choice: to follow the actions our bodies are asking for, or to ignore those feelings for other goals. Sometimes there's a good reason to push on through tiredness, but if you constantly ignore your body, eventually your body will rebel and force the issue. The urge to take 15 minutes in the middle of the day is more often than not an urge to let your mind flow into an abstract dream space after spending too much time focusing. Just a few minutes are all you need.

Naps can be of just about any duration, but for reasons we'll get into later, shorter naps, around 20 minutes, are ideal. Longer naps stretching between 30 and 90 minutes can be counterproductive because of the way that our brains descend into ever-deeper stages of sleep over time. It's difficult to immediately come back to alertness if you've been asleep for an hour; in fact, you might remember a time when you thought that a quick nap would rejuvenate you but instead found that coming back to consciousness was akin to surfacing from underwater.

The time that it takes to come back to awareness is what sleep scientists call "sleep inertia," a metaphor that borrows from an outside discipline—this time, physics. Just as it's hard to fall back asleep after you've already begun your morning routine, it's difficult to wake up once you've already started sleeping. The deepest sleep stages correspond with the most abstract types of thought that we typically ascribe to the most intense dreams—where the laws of physics and causality seem to fly out the window. Your brain needs time making the transition away from dream-physics which makes coming back to the physical world mentally challenging.

Some people will say that they're "bad at napping" because they find it difficult to shut down their habitual and constant daytime alertness. They lie down on a couch or in their bed with their eyes closed, and rue the reality that they don't fall into proper sleep. They declare the nap a failure if their consciousness doesn't bleed away for at least part of their planned downtime.

These people often ask me if there's a trick or a hack that will give them the "proper" sleep. What they don't realize is that even if they don't descend into delicious soporific somnolence, the mere act of trying to unwind is still incredibly beneficial. This is the beauty of reconceiving consciousness as a continuum. A break to lie down and rest sets an intention to let the mind drift into abstraction and make novel insights about the world without having to fully commit. A nap can signal the metabolic and hormonal systems to pause the constant drive for alert focus and relax enough to catch their breath. Not surprisingly, when someone lets go of the goal of falling asleep, they often actually do drift off.

Beyond the health benefits and creativity boosts, it's important to remember that sleep is also often intensely pleasurable. There are few things I enjoy more than winding down and going to sleep at night, aware that I will soon be dreaming.

Over the course of this book, I'm going to document what we know about dreaming, offer you techniques to get the most out of your short rest, and spell out how dreaming will change your life. But more than any of that, the reason I am writing it is to give you the perfect excuse to put down what you're doing and get much-needed and much-deserved rest. If you tell your co-worker, spouse, boss or child that you need to take some time for yourself to put your head down for a few minutes, I will have done my job.

Included in these pages are scientific findings and anthropological observations that will peel back centuries of conditioning that put dreaming in consciousness's back seat. In the process of reporting, I designed and executed an experiment that gave me new insights into what it means to look for efficiency in our sleep cycles. It is also a historical journey into popular sleep and relaxation literature dating back to the 1800s, where insights into rest and recovery have been repeatedly discovered and forgotten time and time again. Along the way, I will give you tips and tricks that have made me an effective and enthusiastic dreamer.

If I'm going to be completely honest, I didn't start out with the intention of digging deep into the dream world. Indeed, this was supposed to be a book about napping. Those sweet, short breaks in the middle of the day restore your dwindling spirits, boost creativity, foster emotional resilience, and strengthen the immune system. The book I would have written would have proclaimed that good sleepers live longer and that they even tend to make more money. The book might even have sold well. There are only two or three worthwhile reads on the topic, and they frequently lurch into overly technical details that offer up complex algorithms for ideal nap timing and duration. The world could do with a great book on napping, and though I have not actually written it, I can offer you a quick summary:

You work too hard. Lay your head down when you start to feel tired in the middle of the day, and aim for a duration of 20 minutes or 90 minutes (but, critically, not in between) so you can avoid waking up feeling groggy and lethargic.

I would have given you that message in a dozen different ways to tell you what you probably already knew: Naps are healthy, restorative, and will make you a happier and better adjusted person. If you are like everyone else who grew up in the modern, industrialized world, you are probably not napping enough.

So there you are. You can consider yourself up to speed on the topic. Go take a damned nap, already.

The reason that this is not a book on napping is that when I tried to devise ever more efficient napping techniques, I found that it was almost impossible to write about napping without also tackling the questions around what it means to dream. This was strange to me, because before researching this book, I thought that dreams only really happened during the most sound and prolonged sleep stages. I didn't quite realize yet that we're dreaming all the time. Even while we are awake.

I accidentally plunged myself headlong into a much

more interesting and fertile topic about consciousness writ large. You see, when a person is successful at remaining aware of the transition between wakefulness and sleeping, they gain a fundamental insight into the way the brain processes information. The dream world is, in fact, a world unto itself, with its own physics, underlying processes and evolutionary foundations that, as far as your brain is concerned, is as real as the material reality that has you reading a book about dreaming. It's a world in the same way your house is a world—suffused with meaning, emotions and associations. It's pregnant with wisdom, and even though you probably won't remember everything that happens in it, what you do in your dreams forms the emotional core of just about everything you experience in life.

When I envisioned all of the different directions that I could take this project when I began diving into the strange world of dreaming, napping, darkness and human evolution, I imagined I would spend a good amount of time bumping around different sleep labs and recording the amazing routines of top athletes, CEOs and visionary thinkers. Many of them, after all, dedicate a good portion of their day to recovery and free thinking. Of course, versions of that book already exist. Matthew Walker's *Why We Sleep*, Satchin Panda's *The Circadian Code*, and Sara Mednick's *Take a Nap! Change Your Life* are all exemplary specimens of the genre. Meanwhile, Tricia Hersey situates napping as a revolutionary act for black liberation and as an antidote to grind culture in her book *Rest Is Resistance*. Other books take different tacks, yet all come to the same basic conclusion: The world we live in is incredibly hostile to sleep. Work patterns, life obligations, diet, historic oppression, kids, drug habits and electric lighting all play a role in keeping our eyes open when we should instead be trying to keep them shut.

After a year of studying napping, I realized that my real interest wasn't in finding a napping formula that would make me more efficient at some real-world task. Instead, I wanted to uncover the continuous links between my subconscious and everyday activities. I wanted to understand how who I was in my dreams was also who I am all day long. I learned that getting good at falling asleep also meant getting good at being awake. I learned to cross sleep's threshold while keeping a part of my consciousness alert to the transition. I studied both ancient rituals and cutting-edge technology that made it possible to actually influence the content of dreams. This wasn't the lucid dreaming of the 1970s—a place where the sleeper is aware that they are sleeping and can then take advantage of the dream world—but an awareness that the experience of dreaming actually forms the foundation of human emotions. I learned that I could influence my emotional baseline by seeding ideas into my mind as I fell asleep. Naps were simply one method to manage that transition. There were two others, as well: the sleep that we fall into at night at our usual time, and the second sleep that happens to just about all of us when we wake up in the middle of the night and ponder questions that we don't have time for during the day.

Whether or not you actually read any more of this book, I want you to know one thing: Dreaming is central to who you are. Dreams determine how you show up in your waking life, they refine your emotional bandwidth, and they are vitally important partially because you don't remember them. More than that, I also want you to know that the dream world is every bit as real as the real world except that it behaves in a way that you would have never thought possible.

Sleep played a vital role at every stage of evolution, dating back to a time when single-celled creatures modulated their metabolism in relation to day and night cycles. Animals dream as surely as fetuses do. The very first written records in human history also deal with the interpretation of dreams. In the next chapter, we will examine both the anthropology of dream states and how darkness alters the human mind.

2

EVOLUTION, DARKNESS, AND DREAMING

As far as we can determine from the fossil record, the first primates sprang into existence about 55 million years ago and spent the majority of their lives inside the safety of hollowed-out tree trunks. They huddled with their relatives and slept an estimated 17 hours a day, only venturing outside under the cover of darkness to forage for what little food they could scavenge. From an evolutionary perspective, if you get eaten before you pass on your genes, your species doesn't have much of a chance. So sleep poses a peculiar challenge. Unconscious animals are incredibly vulnerable—and they're also not spending time locating their next meal or finding new mates. As the late, great sleep researcher Allan Rectschaffen once wrote, "If sleep does not serve an absolute vital function, then it is the biggest mistake the evolutionary process ever made."

Modern humans sleep less than half as much as our arboreal ancestors, but anthropologists conducting careful historical and cross-cultural sleep studies have discovered that the secret to our success in spreading our genes and even taking over the globe might be intimately tied to the way we sleep. It's not just the duration of our downtime. Our real advantage comes from our incredibly flexible sleeping abilities, which can change with our environment and ecological circumstances. We can thrive near the poles where the sun hangs above our heads for nearly 24 hours straight, or the other half of the year, when it plunges into darkness. We can sleep to avoid the heat or wake up to catch the warmest part of the day. And we can adjust our sleep schedules according to whatever time is best to avoid predation. Not every animal is as flexible.

As Rectschaffen commented, all that time in bed comes at an obvious cost—every hour asleep comes at the expense of acting in the world. So the benefits of taking a few hours off must ultimately outweigh the costs, or our species simply wouldn't have devoted so much time to perfecting it. Mammals—and humans in particular—thrived because of our ability to adapt to the changing environment with novel thinking. Bigger brains correlate closely with an increased amount of time spent in Rapid Eye Movement (REM) sleep, which, as we will see in detail in the next chapter, confers a remarkably efficient ability to form novel insights into the world around us and help solve problems that we might not have known were out there. Zoologists have also shown that mammals with larger brains correlate with higher white blood cell counts and fewer parasites overall—an indication that consciousness sets us on an evolutionarily successful path.

As human ancestors got larger and smarter, they spent more time on the ground, where they were even more vulnerable to attack than their squirrel-sized predecessors. Cognition alone was not enough to protect them. Instead, early humans slept in groups and used sentinels to watch for threats so the rest of the group could enjoy the benefit of restorative sleep. Recent studies on a hunter-gathering group in Tanzania called the Hadza and highland tribes in New Guinea show that indigenous people almost never go to sleep at the same time. Instead, sleep is a social activity that revolves through the group, with some people keeping watch while others slip into dream time. This segmented sleep happens for most people in two distinct phases in the night, with a bump of activity in between that the historian A. Roger Ekirch calls "second sleep" in his book *At Day's Close: Night in Times Past*. A third phase of natural sleep typically took the form of afternoon naps.

Anthropologists call this a "polyphasic" pattern, and as far as we can tell through the power of ethnographic analysis, these sorts of segmented sleep cycles are as close to a human's natural sleep as we can estimate. The anthropologist David Samson argues that communal sleep cycles helped form the basis of the first human social groups. Sleeping in shifts built trust among the group while it entailed the sort of communication and coordination that eventually translated into effective hunting strategies during waking hours.

While we were becoming more efficient sleepers, when it came to managing our sleep cycles, humans also began sleeping less and less. The seven to eight hours that the average *Homo sapiens* needs to act in the world is significantly less than the sleep time of every other primate species. This means that we must have particularly efficient physiological mechanisms that accomplish the same amount of sleep work in less time than our arboreal relatives. The key might be the greater proportion of our sleep cycles that we devote to REM. More efficient sleep lets us also adapt our sleeping patterns to whatever ecological context we happen to find ourselves in. This is an important advantage. Yet as adaptable as we are, there are limits to that flexibility.

Barring the invention of a time machine, sleep researchers aren't ever going to be able to measure human sleep cycles over evolutionary time scales. But one famous study, conducted more than a century ago, revealed something dire about the changing nature of sleep in the modern world. The psychiatrist Stanley Coren wrote that in 1910, an epidemiological survey recorded the sleep habits of the average American. This turned out to be a pivotal baseline in American history that just happened to come out three years before Thomas Edison released the first tungsten filament light bulb. Until that date, light bulbs were expensive vacuum-pumped affairs that gradually grew dimmer over time. Edison's invention radically altered the way we lit our world, almost as quickly as flipping a switch. The baseline study that came out before the tungsten bulb reported that ordinary people slept nine full hours every night. The electric light bulb removed a full hour and a half of sleep from every living person on Earth. That's 547 fewer hours of sleep per year. More than 45,000 lost hours of sleep over a lifetime.

In our distant past, humans would make up lost sleep hours at night with restful periods in the middle of the day, the natural ebb and flow, but no Edisonian invention popped up to promote naps. Instead, increased productivity came at the expense of total sleep.

Edison made no bones about his distaste for wasted time in bed. He claimed he never needed more than four or five hours of rest a night (which, if true, would have made him a statistical outlier for healthy sleepers). His light bulb and cheaply available electricity kick-started a new era of global productivity. More light and less sleep meant that there was more time to actually do things. In his biography, Edison wrote that he was sure that it made people smarter, too. (Maybe this is where we get the phrase "a light bulb went on" in someone's head.) On a drive through Switzerland, in a time when the country was expanding its electrical grid for the first time, he wrote, "I visited little towns and villages, I noted the effect of artificial light on the inhabitants. Where water power and electric light had been developed, everyone seemed normally intelligent. Where these appliances did not exist, and the natives went to bed with the chickens, staying there till daylight, they were far less intelligent."

Edison's gusto certainly served his own bottom line. This was also the view of just about everyone who could get their hands on an electric bulb. Almost every paper and government broadcast in the last century heralded this moment as the beginning of the modern era—when we threw off the chains of nature and claimed a brave new future driven by rational thought, hard work, and the dazzling capabilities of our inventions.

But in the immortal words of the fictional scientists about to bring dinosaurs back from extinction in Jurassic Park: "Your scientists were so preoccupied with whether or not they could, they didn't stop to think if they should." The sudden appearance of cheap electric lighting broke the human race away from one of its most stable natural relationships. The circadian rhythm sets its clock based on the available light given off by the sun and rotation of the planet. Now, literally, almost the entire human race sets its own internal clock based on the burning of fossil fuels.

Imagine what it must have been like to live back in 1914. Every home would have an electric light bulb or two, and maybe even a radio to hear the national broadcast from the president or some of the musical precursors to what would eventually become rock and roll. If you were fortunate enough, you or your neighbor might even have access to an automobile. Meanwhile, your Luddite parents and grandparents spent their childhoods studying by candlelight and probably only had one or two books (most likely the Bible) to keep them company. Yours was a fast-paced new age of radical transformation. But to the eyes of a descendant living 120 or so years later, 1914 would have been almost unimaginably slow. Now, in the age of artificial intelligence and internet-powered home food delivery, the pace of human life grows faster almost by the minute.

The decline in total sleep—both at night and as naps during the day—probably came on too quickly for us to evolve a way to compensate for it with a more efficient brain process that divides us from other primates.

Still, in 1990, a specialist in psychobiology at the National Institute of Mental Health named Thomas Wehr wanted to find out if the sudden electric illumination was sparking evolutionary changes or if humans would revert to evolutionary sleep in its absence. The experiment was straightforward: He'd send people "back in time" by restricting their access to illumination to when the sun was actually up. He took out every bulb, luminescent clock and glowing computer screen from their homes. This kept his subjects in darkness from dusk until dawn, only allowing them to have light in a natural rhythm. For the first three weeks, the subjects slept moderately more than normal—adding about an hour of sleep a day to their routines. But in the fourth week, something magical happened. Their total sleep remained the same, but the sleep patterns spontaneously divided into two segments of four hours each with a two-hour "quiet rest" phase in between. For anyone who likes staying up late by the light of a TV screen or bedside lamp, the results should make you stand up and pay attention, because we're staying awake on what amounts to borrowed time.

Not only that, but the natural gap of semi-alertness is a unique state of consciousness between total awareness and dreaming that all of our ancestors would have experienced every day but that modern denizens of eternal light miss out on. Clark Strand, former editor of the Buddhist magazine Tricycle and author of the book *Waking Up to the Dark*, wrote that this gap in nighttime sleep could be a gateway to meditation and prayer precisely because it seamlessly blends the experience of dreaming and being awake. Strand writes that during his time in a monastery, he and the other monks found praying in those nighttime gaps led to elevated spiritual experiences, and they took to calling it the "Hour of God."

Meanwhile, Wehr discovered that the people in his study had elevated levels of the hormone prolactin starting shortly after dusk and then staying almost twice as high as normal through the entire eight hours. Prolactin levels in electrically lit households decline steadily over the course of an evening, bottoming out just before dawn. Prolactin isn't the sort of name-checked hormone that comes up in biohacking podcasts in the same way that dopamine and serotonin do, but it plays a vital role in human health. The pituitary gland secretes prolactin during pregnancy and after birth. It helps manage at least 300 identifiable functions in the body, from metabolism and immune function to mood and behavior. It also plays a central role in milk production and tempers the effects of both estrogen and testosterone. Birds secrete prolactin to keep them calm while they incubate their eggs. Its central role in helping both newborns and mothers bond has made some people call it the "attachment hormone."

Strand writes, "It creates a feeling of quiet, security and peace."

Wehr discovered that turning off the lights and keeping them off returned his subjects home to a protected space in their own consciousness that their hyper-lit colleagues never experienced. When Edison began his mission to banish the darkness from human life so that he could supercharge overall productivity, he also amputated a state of consciousness that was fundamentally restorative for nearly the entire Earth-bound population.

Wehr later wrote: "This is a state of consciousness not terribly familiar to modern sleepers. ... Perhaps those who meditate today are seeking a state of consciousness that our ancestors would have considered their birthright, a nightly occurrence." Given this, it's no surprise that just about every religion founded more than two centuries ago has a tradition of midnight prayer or meditation that coincides with the timing of a second sleep. In Islam, it's called the Tahajjud, while in Judaism it's Tikkun Chatzot—both of which roughly translate to "nighttime prayer." In Christianity, Catholic monks of the Carthusian order practice a "night office," whereas in Hinduism, Yoga Nidra, or "yogic sleep," may reach back all the way to the Mandukya Upanishad in the second century A.D. Meanwhile, the Taiseki-ji temple in Japan has held a midnight service called Ushitora Gongyo every night for the last 800 years.

Most of us who grew up in the electrically omnipresent modern world feel uncomfortable with the prognostications of various religious orders. We have seen how moribund theological thinking can become and how faith in a higher power is often an antithesis of rationality. Be that as it may, I have no doubt that you have noticed how your own thoughts change if you happen to wake up in the middle of the night and look at your ceiling in total darkness. For several years, I've enjoyed a bedtime between 9:30 and 10:00 p.m. and have noticed that I will often wake up briefly shortly after midnight. While I'm not religious in any organized sense, in those moments I may mull over the ebbing memories of a recent dream and then lie awake long enough to allow my thoughts to drift . . . well . . . just about anywhere. I consider my everyday worries at a slower pace. I feel safe in the darkness and in my bed, and enjoy the time with my thoughts. Eventually that gives way back into uninterrupted sleep. An ancestor or particularly religious version of myself may have considered that the hour of god. But for me, it is evidence of a sort of evolutionary pulsation in the ordinary fabric of life itself. It is not wasted time, but rather a time that our species must have always made use of. It is as natural as the way that all biological creatures on earth sync their own biologies to the rotation of the sun.

Second sleep is a sort of reverse nap. It's a natural oscillation between alertness and dreaming. Where napping might replenish flagging spirits, the second sleep lets awareness ruminate on dream time that you might otherwise forget.

Almost no creature on the planet maintains a constant and steady level of activity from dawn until dusk. Most animals favor certain times of the day to gather food, locate potential mates and secure housing. The vast majority of species are crepuscular—meaning they are most active at dusk and dawn in the twilight between light and darkness. Animals spend the rest of the non-active time in a much lower state of excitement, tending to their burrows and nests, grooming and ruminating while their minds likely wander in what most of us would consider daydreaming.

Even if you aren't particularly attuned to the ebb and flow of the natural world, you've no doubt noticed the chorus of bird, frog and insect songs in the morning and the tendency for larger wild animals to appear as the sun dips below the horizon. Certain apex predators will hunt in the full light of the day, but, especially in warmer climates, almost everything comes to a standstill when the sun is directly overhead.

In 1770, the Georgian author Charles Burney wrote, "Only Dogs and Englishmen are seen out of doors at noon, all else lie down in the middle of the day." The quip started as a dig at Colonial busybodies pursuing an imperial agenda, but the saying survives and resurfaces again and again throughout history, as it points at the relative insanity of working non-stop throughout the day—ultimately taken up as the chorus to a popular song in the 1930s. The image of a sunburned man peeking out from behind a pith helmet is frankly absurd. But it's also an absurdity that most everyone in the workaday world unabashedly imitates. Those of us who ignore the natural urge to nap are just as unnatural and out of place with our biology as the mad dogs who roam the streets without a connection to the natural order of the universe.

I don't know how practical it is, at the end of the day, to truly embrace a polyphasic lifestyle. While I can imagine a utopian world where sleep patterns track with the sun and no outside perturbations or responsibilities interrupt what our bodies truly crave, I'm enough of a realist to understand that it isn't likely to happen. Even so, there's no reason we can't push back against the implied obligations. There's no reason we can't think of creative ways to circumvent the grind culture that keeps us apart from the natural world—even if we don't achieve a total revolution.

We can start by admitting that it's okay sometimes to ponder the universe projected on our bedroom ceilings at night and not be overly concerned that it will undermine our overall efficiency. We can realize that the varied states of consciousness that we undergo throughout the day have been vital to our survival over the eons. But what does that consciousness really look like? For that matter, which came first: dreaming or waking?

3

DREAM STATES

What is your earliest memory? Mine happens to be falling down a set of stairs with my arms out in front of me, as if I were launching into the air like Superman. I can remember the texture of the unfinished wood of the stairs, the railing that my hand failed to grasp, and that my mother was only about ten feet away in the kitchen. The only problem with this memory is that I have no idea if it actually happened the way my three-year-old brain encoded it. At the very least, a fall down an entire flight of stairs at that age would probably have left some lasting marks. Maybe my first memory was actually just a dream. Several of my most distant recollections are similarly suspicious.

If I didn't make it up, another vivid memory from around the same time possibly occurred in a blue preschool building where a teacher was trying unsuccessfully to get a class of unruly students to quiet down. I could see her appeal to the class wasn't working, so I decided to help out by yelling out in my loudest voice for everyone to "SHUT UP!" That worked, but the teacher was furious at the vulgarity. She started to run after me for what I knew would be a terrible punishment. I sought refuge in a papier-mâché tree fort in the corner of the room, but she dragged me out and took me to the bathroom, holding me over the toilet and telling me to say the words again. If I did, she'd flush the toilet and I'd never be able to speak again.

My brain is pretty sure that both of these events definitely happened, but my rational mind is suspicious. The stories are emotionally charged, frankly terrifying, and yet somehow also contain a kernel of potentially useful information: Be careful on stairs, and don't piss off your teachers. Although they're my very earliest memories, they also contain other information—the layout of my home and preschool buildings, relationships to various people and places, and even an awareness of social norms—all of which I must have picked up at now-forgotten earlier times.

Now I view these earliest memories as the blurred line between dreaming and wakefulness, memory, the process of forgetting and even the nature of consciousness itself. It's easy to confuse the relationship between memory and consciousness. It's natural to think of who we are in terms of the things that we have done in the past and assume that there's a continuous story that connects our most recent experiences to our very origins. Yet how can that be true if very few of us have clear memories of even a few days ago, let alone events that occurred decades in the past? Still, we are pretty sure that we were conscious in those times that we don't remember. My first memories hint at awareness of prior events. So who am I, anyway? Am I my memory? Or am I just a memory's beneficiary? More to the point, if memory is not the root of consciousness, then what is? When do humans first become conscious, anyway?

A theologian might tackle that last question by

suggesting that life begins anywhere from the moment of conception to sometime after birth. Meanwhile, in her book *Death Without Weeping*, the anthropologist Nancy Scheper-Hughes reported that people in places with high infant mortality, like the favelas of Brazil, contend that life doesn't begin until up to three years after birth, when they officially bestow a name upon a child. Regardless of whatever arbitrary line we want to draw, most people would agree that infants are conscious even though they don't form long-term memories. So let's go back even further.

While it's impossible to know for sure exactly what a fetus experiences inside of a mother's womb, the best data we have suggests that it resembles something pretty close to what we would call dreaming. A human fetus transitions through three stages of brain activity in the first two trimesters that closely resemble REM and NREM sleep, as well as an intermediary stage that has characteristics of both.

Then, in the final trimester, the fetus begins to wake up. It kicks, bounces and explores the tight confines for two to three hours a day. As Matthew Walker notes in Why We Sleep, in the final weeks before birth, the fetus will also spend almost nine full hours a day in REM sleep, increasing all the way to 12 hours just one week before it enters the world. Walker writes that "REM sleep acts as electrical fertilizer during this critical phase of life. Dazzling bursts of electrical activity during REM sleep stimulate the lush growth of neural pathways all over the developing brain." This process, called synaptogenesis, links the brain together into a consciousness-generating machine. If those neural firings are anything like the same sort of brain activity in a child's or adult's brain, then we have to assume that the fetus is experiencing something in the womb—and that something is most related to what we think of as dreaming. In this sense, every human's most primordial experience of being alive is a dream. Or, to push that logic just a little bit further, who we are when we are dreaming is more essential than who we are when we're awake. The only catch is that at the point when they're still in the womb, fetuses haven't really experienced all that much.

Mark Blumberg, a neuroscientist at the University of Iowa, has spent a lot of time chewing on the question of fetal dreaming, and tried to understand it through the characteristic twitching that we all do when we sleep. While someone might look at a sleeping dog that moves its feet and assume that their twitches must relate to a dream of chasing a dream rabbit, Blumberg comes to the opposite conclusion. In a series of papers, he suggests that REM initiates random muscle twinges as a way for the brain to start learning the body that it will (at least for a fetus) one day command. It's an interesting idea, but it doesn't answer anything about the content of those dreams. What does dreaming about muscle twitches feel like?

I asked a different neuroscientist, this one at MIT, what he thought fetuses dream about, and he asked me if I thought humans ever truly wake up. Maybe, he posited, being awake is just an extension of what happens when we dream. If our experience of consciousness occurs primarily in the brain, then what we sense of the outside world is not a direct experience of objective reality. Everything we feel and detect in the environment passes through our peripheral nervous system as data that our brain assembles into a cogent model of the world at large. As Blumberg notes, the brain has to learn the body, but it will always be just a brain locked inside a case of flesh. The signals it detects from the world through the nerves is known as "perception." If consciousness is in the brain, then its truest form occurs when it's not trying to make sense of the outside world, but rather when it contemplates itself. Starting from a time even before we're born, true consciousness happens in our dreams.

So what, exactly, happens when we dream? I've already mentioned a few of the most important sleep stages, but understanding how one stage bleeds into another will shed light on how impressively active sleep really is.

As a person begins to relax and fall asleep, the mind slowly peels back in layers of relaxation. That state between waking and sleeping is called hypnagogia, where focus in the waking world runs up against the shores of dreaming. It's a place where abstraction meets focus in a dazzlingly useful way.

Hypnogogia occurs in the first of the four main sleep stages—during the transition between wakefulness and true sleep. It's a liminal space where daydreams transform into dreams in an inner-tidal zone between alertness and abstraction. Most people wouldn't call hypnagogia true sleep. The dreams in hypnagogia aren't quite as robust as in the later stages. After four or five minutes in stage 1 hypnagogic sleep, the characteristic brain waves, heart rate and body temperature drop leads into stage 2 sleep, where your brain starts to consolidate and process motor learning and muscle movements to transfer them into long-term storage. When you're learning a new activity like dancing (or, more specifically, in terms of what most sleep scientists directly study, if you are a rat practicing a maze), stage 2 sleep refines those types of skills.

Where you might wake up from an errant cough in the room during stage 2 sleep, deep sleep (stage 3) brain waves take on a languorous pace, breathing turns shallow and regular, and the brain stops producing the stress hormone cortisol. In deep sleep, the body focuses on regenerating tissues, restoring organs and boosting the immune system. It is the realm of the unconscious handyman. If you wake up from deep sleep—a hard task in itself—you might feel like you are underwater. Any dreams that you remember will probably seem rather ordinary, almost linear. Maybe you were dreaming about starting up your car or getting a bite to eat. The major changes that happen in deep sleep focus on bodily processes rather than mental ones. Yet there is still a deeper level to move to. In stage 4 sleep, memories begin to alter and uncouple while the body works on restoring the cellular infrastructure.

That all changes soon with a sleep phase that you've already heard something about: REM sleep, where the eyes dart back and forth and dreams come alive. Interestingly enough, the body doesn't descend straight into REM after deep sleep. Instead, it detours back to stage 2 sleep, as if it's resetting its footing after focusing on repair of physical tissues in order to begin an entirely different process. After about eight minutes after that reset—or roughly 80 minutes after you first closed your eyes for the night—REM sleep sets off a fireworks show.

During REM, your heart ramps up along with your blood pressure and your body temperature plummets. The brain sends its paralyzing signal to your muscles so that you can't start acting out your dreams in real life. Then begins a free association of pure and unadulterated creativity. As the brain cycles through whatever got collected in your bin of short-term memories from the day, a cognitive process kicks in to discard useless information and forge them into gists of memories that later turn into the basis for your emotions. Another way to say that is that, as the brain churns through the grist of short-term memories, it spits out two byproducts: emotions and long-term memories. All of this plays in front of your sleeping (and non-remembering) eyes like video speeding forward and backward on a reel of film. If you spend your day learning a new language or picking up a new skill, REM sleep processes that information and stores it. If you happen to miss out on REM sleep during that time, you will make significantly less progress than you would have liked.

Every night, you will typically go through three full sleep cycles, moving in and out of the various stages of sleep in more or less the same order on repeat until you wake up in the morning. Researchers have meticulously categorized and measured physiological changes in each state so well that now even consumer devices that attach to the wrist or finger can reliably detect the transition between them.

It's worth noting that most of what we know about the different sleep stages comes from external measurements—from changes in the pulse rate and electrical conductivity. All we know about the content of sleep is from the answers that groggy participants fill out on questionnaires or from seeing how a person's performance changes in specific tasks when one or another sleep stage gets interrupted. Despite decades of laboratory research, there is still no way to directly study the subjective experience of being asleep. Instead, we have to rely on what freshly awakened sleepers can remember.

Relying on fading memories is tricky, as I noted in the beginning of this chapter. People don't generally retain memories of their formative childhood experiences, even though having those experiences is vital to their development. Dreams fade within seconds of opening your eyes in the morning. You might ask: What possible value is there to an experience you typically can't remember? The answer might surprise you: It's often more useful to forget things than retain them.

Dreams aren't important for their narrative content, but for their function in storing and processing memories. Just like a computer hard drive, the brain has a limited amount of storage space. If we remembered everything that our senses collected in a day, we would soon be overloaded with irrelevant information. There's no point in remembering the specific color of a houseplant, your spouse's request a few days ago to empty the dishwasher, or the sound of a particular motorcycle revving its engine a few blocks from your home for the rest of eternity. The brain needs a way to condense and synthesize the massive amount of environmental data that you collect all day long and reduce it all to something useful. So it stores all of that information into short-term memory, which you can think of as RAM on a computer, in order to delete the unnecessary bits in your downtime. Dreams sift through those short-term memories and discard the vast majority of the information while condensing relevant information into gists. Those gists are the basis of your emotional life. Dreams forge emotions.

This may seem weird. Emotions are feelings that we have about the things in the world around us. Sometimes they feel irrational, even random—as in "I don't know why I don't like the color yellow. All I know is that I don't." Other times, emotions hold vital information about how to act in the world—such as looking down over the edge of a great height and feeling an overwhelming desire to get back from the edge.

It turns out that even though we don't think about where they come from all that much, emotions are remarkably effective algorithms by which to navigate the world. Many of us live under the illusion that our actions are based on logic, not emotions. To test that idea, I'd ask you to use your logical brain to think about your own actions throughout the day. Do you want to eat a delicious plate of food because you know that its ingredients will give you the optimal nutrients to power your day, or do you have an emotional attachment to the taste you think it will have? Do you engage in conversation with your friends because you desperately need specific information from them, or because you simply like the act of conversing? Do you look at a bathtub full of ice and think that you don't want to jump in because it just doesn't feel right, or because you've seen the data and know for sure that it's a bad idea? While we generally think we have good reasons for the things we want to do, more often than not, we make decisions because of our emotional responses. Those responses all emerge out of the gists of dreams.

If you try to recount the plot of a dream, it often doesn't make a lot of narrative sense. One minute you might be in a room talking to your mother, and the next you are trying to build a treehouse out of spatulas, only to find out that secret agents are trailing your every move and, oh, my God, you're late for your flight! If a sleep scientist recorded the activity in your brain at the time of this specific dream, they would find that most of the activity occurred in the emotional processing areas of the hippocampus. In other words, instead of trying to understand dreams in terms of their plot, try to understand how they work on an emotional level. Seen in this way, the brain is actually combining events in your waking life with different emotional situations, feelings and responses. This emotional processing goes on to form the bedrock of how you feel about the world in general.

Perhaps you've noticed that if you don't get a good night's sleep, your emotional reactions the next day are a little off. Maybe your perception of what other people are saying is unusually negative. Have you ever responded to an email or a tweet with an uncharacteristically harsh tone? This could have happened because you failed to give your brain enough time to process the emotional content of the day before. This point is important, because it connects dreaming to activity in the real world.

One way to think of consciousness is that it's something like a computer program running on the hardware of the brain. Where the ones and zeros make up the bits and bytes of computer programs, the bits and bytes of human consciousness are a combination of sensory data from the outside world and emotional reactions. To briefly summarize the process of cognition, every sensation we feel first has to come in through the peripheral nervous system and the spinal cord and then enter the brain in its lowest levels in an area known as the limbic system. You can think of the limbic system as a sort of library of all prior sensations. At first, that sensory information is only data that carries no emotional value, and it's the limbic system's job to catalog and compare the raw data against a library of prior sensations to determine its emotional value.

If, however, a brain has never experienced a particular sensation, the limbic librarian reroutes the signal to the paralimbic system (think of this part of the brain as a book binder). The book binder takes the new sensory data and assigns it to the brain's current emotional state and then returns this new book back down to the limbic library for filing. This new book is what I call a "neural symbol," and it's the basis for all cognition. The next time that a person detects that same sensation from the outside world, the limbic librarian does not need to go through the hassle of creating a new neural symbol, but instead pulls the old one off the shelf and enters it into the stream of cognition. While one neural symbol doesn't convey a whole lot of information on its own, just like a computer running a program composed of ones and zeros, consciousness needs hundreds, thousands or even millions of neural symbols to process a single thought. The important thing to understand about this process is that, at its most fundamental level, the way you understand the world is almost always rooted in your emotional past. This is the central process that I wrote about in *The Wedge* to describe methods that intentionally change the way we feel in stressful environments and situations in order to gain a certain amount of control over our autonomic bodily processes.

The Wedge helps build resilience that will allow us to control many physiological and psychological reactions in the waking world, but I've often felt that I failed to truly explain how emotions form in the first place. The example of the limbic librarian and book binder assumed that emotions just sort of arose out of the psychological ether— and I wasn't able to adequately explain their origins. At least not until I delved into the true function of dreaming.

This made all the more sense when I learned that when we dream, the brain suppresses physical movement of your limbs so that you don't get up and start moving around the waking world while you are asleep. To help do that it also almost completely stops secreting noradrenaline, a key stress hormone. This means that the content of dreams occurs in an environment that is almost chemically immune to stress—all the while reliving relevant emotional experiences and creating the neural foundations for interpreting real-world sensory data in the future. Dreams are a safe space in which the brain works out emotional problems without real-world consequences.

If you don't devote enough time to that emotional processing, as in the case of a poor night's sleep, unprocessed negative impressions only gather force in your waking life, as you are unable to discard the irrelevant information to create the big-picture emotional takeaways. Walker's book contains a much more thorough description (p. 206).

More than just emotions, dreams also affect our ability to focus and take on creative projects. Waking life often requires the ability to focus on single individual tasks for extended amounts of time. Whether you are a modern office worker composing an email or a stone-age hunter crafting an arrowhead out of flint, focusing has obvious benefits to accomplishing specific tasks. Alert minds are intrinsically narrowly focused. Take an experiment using word associations, for example.

If you show a fully alert test subject a series of flash cards with common words on them and ask them to say the first word that comes to mind, the associations are almost always closely related. For example, if a flash card presents the word "fork," an alert respondent will likely respond with a word along the lines of "food," "knife" or "spear." In each of those cases, there's an obvious relationship between the utensil and related objects.

However, if you perform the same test on a person who has just woken up from REM sleep and offer them the same word, they will respond with increasingly abstract and non-linear associations. They are far more likely to say "hippopotamus," "aircraft" or "typewriter" instead of any word that their alert brains would come up with. The sleeping brain prefers abstraction to focus, and indeed cannot even process closely associated thoughts.

Human consciousness needs these two polar responses in order to make sense of the world it inhabits. Focus allows us to home in on and accomplish specific tasks, while abstractions allow us to make general insights about the world and then discard irrelevant data. Those abstractions can lead to the surprising discoveries that underpin human creativity.

We can describe the difference between abstraction and focus by considering the brain's neural circuitry. If a thought existed on a single connection between two neurons (a simplification, for sure), then a focused association likely looks like the most direct path from one neural location to another. An abstract association, on the other hand, takes a longer path, touring around and perhaps even stopping in an entirely different side of the brain before deciding on an ultimate destination. Imagine a child wandering around inside a museum, moving from ancient Egyptian artifacts to the modern paintings of Jackson Pollock and onward to medieval Chinese pottery. The child would come away with an entirely different experience than someone who followed the museum's carefully pre-planned tour directions. This process of collecting different (and possibly random) data allows for novel connections, paths and creative associations. While many of those associations will have no real-world applications, abstract thinking often generates more efficient solutions than a focused brain would have ever come up with.

Have you ever gotten frustrated with a problem and taken a step away from whatever it is you were doing, hoping that a solution would come to you from the act of getting away? Maybe you let your eyes glaze over or stared at the ceiling or at no place in particular as you turned your thoughts inward and relaxed. These daydreams are the cousins of dreams. They're not too different at a neurological level, either. Great abstract thinkers, artists, scientists, military leaders and political savants often credit their dream life to brilliant discoveries, surprising maneuvers and immortal pieces of art.

Just about everyone daydreams from time to time, but great thinkers occasionally employ more systematic ways to access the creativity of dreaming. Salvadore Dalí and, yes, the inventor of the eclectic lightbulb, Thomas Edison both used what is known as the ball bearing or spoon method to get the most out of their naps. The technique was simple: In order to conjure creative solutions to the problems in front of them or dream up an impossible vision for a painting, they would lie down while holding a heavy ball bearing or a spoon in their hands, and then they would try to fall asleep. Edison was said to put a cast iron pan directly below his hand. As the men drifted off, the tension in their muscles would gradually release, and the object would clatter to the floor or into the pan. The sudden sound would then rouse the sleeper mid-dream, and they would scramble to record their thoughts before the memory of their abstractions faded. As their brains descended into the abstract thinking of the hypnagogia, their minds began making novel connections between real-world events that they might have otherwise missed. By interrupting the natural sleep process, they were able to harvest some of the soon-to-be emotional data into ideas that would have more practical applications.

Both men credited this method with supercharging their abilities to tackle novel problems through abstraction. Dalí conjured up a world of melting clocks and tall ships with butterfly wings for sails. Edison no doubt imagined up the coercive efficiency of an illuminated world. Meanwhile, one-day president Thomas Jefferson imagined the inalienable rights of man that would one day form the basis of the American Bill of Rights. The revolution in new thought began in the men's dream states and only emerged into full expression once they woke up at the sound of clattering metal on the floor.

This works because of the peculiar physiology of sleep. Unfortunately, while a person can wake up and make use of the insights that happen in the first sleep stage, once you descend deeper into the pre-ordained sequence of sleep stages, it becomes much harder to directly retain that information.

Enter the laboratory of Adam Haar Horowitz, where sleep science meets the cutting edge of technology. Haar Horowitz has an advantage that his predecessors of the previous centuries did not: an extensive list of biomarkers, brain spikes and general timing of the various stages of sleep. While he was still a graduate student at MIT, Haar Horowitz realized that it should be possible to automate the task of waking people up when they are in different sleep stages so that they surface long enough to record their dreams. His idea was that those brief bursts of semi-alertness, where someone could respond to pre-recorded prompts, would offer a unique window into both the content of dreams—and potentially a way to incubate specific types of dreams to foster various beneficial outcomes. He posted the blueprints for his invention, the Dormio, to an open-source web board known as GitHub so that anyone could build one and be able to wake up at predictable times. The device can also play audio cues during different sleep stages in order to prepare a person to dream about a specific subject.

Since we know that dreaming has a significant effect on emotional processing, Haar Horowitz reasoned that preparing the way for someone to dream about a specific emotionally charged topic might help with grief, mental disorders, anxiety and, well, just about anything else that someone might like to process. "One thing that we've noted is that, if a grieving person has a dream about someone they just lost, then having a dream about them can bring tremendous relief," he told me.

This isn't just the case with grief, he says. Seeding dreams is potentially a radical upgrade to the ball bearing technique of earlier luminaries; it can lead to better outcomes for people about to undergo surgery, change the way people react to anxiety-inducing stimuli, and even reduce the phantom limb pain in amputees.

I asked him how, exactly, it worked. "It can be as simple as using a program in your web browser that plays your own audio prompts at specific times during your sleep cycle," he says before directing me to a plugin on Chrome that does exactly that—without even the need for sensors. According to one study at Duke of 80 people, as many as 92% had a dream directly related to the theme they seeded during a single nap session.

The hard thing about studying sleep and its benefits is that, while we all agree that sleep is necessary for restoring just about every bodily process, we don't have any real window into the content of dreams. We can see how sleep rejuvenates cells, alters our metabolism and stabilizes fraying emotional resilience. We know that people who don't get enough sleep start to come apart at the seams, undermining all of the benefits that are supposed to accrue over the course of our natural cycles. These things are straightforward. However, anyone who studies sleep from the opposite perspective—using sleep as a way to get more benefits instead of just returning to a baseline of human health—discovers that they need to clear a new bar of evidence for the scientific community to take them seriously. How can we definitely prove that dreaming about a dead relative alleviates grief to a measurable clinical standard? How do we know that a dream about playing soccer will definitely improve athletic performance on the field? They're not impossible questions to set up an experiment for, but it does take the combined effort of a scientific institution forming and testing different hypotheses.

It's the kind of problem that any investigator of the placebo effect ends up running into. When I asked the neuroscientist Otto Muzik at Wayne State University about Haar Horowitz's device, he was skeptical. "There isn't really a point to specifically influencing dreams. The brain will process what it needs to when it needs to. There's no reason to think that we can improve on nature's own processes in that matter." While that is potentially true, it seems to me that it doesn't hurt to at least try.

The body has innate healing mechanisms that will bring a person back to health when they're sick or run down. We know that these processes work over time and that getting lots of rest will, more often than not, fix most conditions. Human interventions by the way of medicine are supposed to work above and beyond the body's own healing abilities. In other words, we go to the doctor because we're not sure that our bodies are up to the task on their own. If the doctor sees us and doesn't prescribe anything other than rest, we often feel a little bit cheated out of the benefit of outside help.

Using sleep as medicine short-circuits the sort of medical relationships that we're accustomed to in the same way that a doctor prescribing an inert sugar pill to help maximize a person's innate placebo response also feels wrong. But the question isn't hard to approach from a purely scientific perspective if researchers are willing to put in the effort. Many conditions obviously respond to pharmaceuticals; ask anyone who has benefited from an antibiotic if they could have just "slept it off." Conditions that have a more subjective element—like pain, mental distress, anxiety or depression—often respond just as well, or at least close, to the effectiveness of the best drugs available on the market. Indeed, every drug tested on human populations has to include the benefits of the placebo effect in their data when talking about how good the pills actually were. There's simply no way to study a medication in isolation of the patient who ultimately has to take it.

All of this is a long way of saying that, whatever the innate healing power of the body is, sleep plays a vital role in it. The experiences that we have when we're dreaming, the different stages of sleep, even the twitching of our limbs as we fall unconscious, all matter. Finding ways to sleep better, deeper and more profoundly must also result in a benefit to our overall health.

I asked Adam Haar Horowitz what sort of people benefited most from the dream implantation methods he was developing and how someone could give themselves over more to the healing power of sleep. "It is exactly the same as the placebo effect. The more suggestible a person is, the more powerful the result," he answered.

A fair amount of literature over the past 50 years of placebo studies has shown that the people who are most prone to suggestion indeed have dramatically higher placebo responses—and thus therapeutic value. That's a little bit of a funny thing to say, though, because the very idea of being "suggestible" sounds a lot like becoming a weaker person. In America, at least, we like our rugged individuals, who buck any attempt at undue influence. Isn't it strange that every attitude toward stubbornness also correlates with worse healing outcomes? Maybe worse sleep, too?

But it shouldn't be strange. Healing doesn't happen by any conscious process. We trust that time and the innate power and wisdom of our bodies will take over the process and manage it far better than we could ever do by actively thinking about what needs to get better. As I discussed in the last chapter, sleep evolved as a trade-off between rejuvenation and activity. Good sleep requires a certain amount of trust and openness to the environment around us. Sleeping makes us vulnerable. We lack any conscious ability to defend ourselves from predators and the things that might come upon us in the night. Sleep requires that we relinquish all control over our bodies and minds and simply trust that things will come out right in the end. Any attempt to subvert that order makes us sleep worse and ultimately makes us unhealthy. So, too, does the question of suggestibility.

If we want to get the most out of sleeping and dreaming, we need to also be open to allowing those processes to change our consciousness. We need to be suggestible and malleable to sleep's power. If we want to get even more out of it than what we might on our own, we need to be open to the power of suggestion so that sleep can hypnotize us into the best possible version of ourselves.

There is one caveat that I'd like to make to this statement, and it's one that will appear time and again in this discussion of sleeping and napping: Human technology can be, and usually is, a double-edged sword. Tech companies that have aggressively tried to capture as much of your attention as possible during your waking hours could work

—and reportedly are working—on ways to influence sleeping minds, too. Haar Horowitz's research at MIT showed that simple audio cues during different sleep stages can reliably influence dreams, even in the absence of sleep sensors. There is no reason that a device that tracks your movements throughout the day or detects the sounds of sleeping at night couldn't also time audio messages to discreetly infiltrate your nighttime sleep. Indeed, according to Haar Horowitz's Ph.D. thesis, sleep tracking and advertising and data gathering will probably be a major frontier for multinational companies.

I can't say if this will truly lead to a dystopian future where somnolent ads play across our vision-scape thirty seconds after we enter into REM sleep, but it's probably a good idea to not be so open to the message of optimizing our sleep with techno-gadgetry that we open the door to something that we'd rather not let in.

Regardless of how our society ultimately decides to handle technology that actively optimizes or commercializes various stages of sleep, it's clearly a frontier that humans have been working on for a long time.

Dreams aren't isolated events that only exist during the moments when your eyes dart rapidly back and forth in their orbits. No matter when you wake someone up from sleep, there is a good chance that they can immediately recount some sort of dream or dream impression. Dreams can occur at every part of the sleep process. Indeed, as Haar Horowitz asked me: "How do you know you're really awake?"

Scientists like to designate artificial separations between things in order to make them easier to study and interpret, but human consciousness has no such limitations. Consciousness is an iterative process that moves between abstract and focused states in a continuous stream that dates back for each of us all the way to the womb. Scientists identify different stages of sleep, but the edges of one stage bleed into the next. Even the line between being awake and being asleep is, to some degree, arbitrary when we realize that the quality of a night's sleep affects behavior during the day. We can also observe that our thoughts as we lie awake in our beds at night have a different quality to them than those that we have during the day while at work or in the car. Consciousness is fluid and highly dependent on context.

Sleeping and dreaming are things that we do and experiences that we feel. From my point of view, there is simply no way to talk about sleeping without also diving into one of the most profound questions that humans can possibly ask: What is consciousness?

Before I began delving into these philosophical questions in the year that I began researching this book, I started by examining something that felt a whole lot simpler. Nearly everyone interested in sleeping also wants to know some sort of technique that aids the transition between wakefulness and somnolence. In pursuit of that goal, I realized I needed to run an experiment.

4

NULL HYPOTHESIS

Everyone wants an easy way to fall asleep. We live in a hectic world plagued by worries, constant over-work, way too much light and late meals. More often than not, we head to bed with more energy than we know what to do with, and all we want is a simple trick, hack or pill to induce unconsciousness. We want our own personal chauffeur into the dream world. Whether or not you consider yourself a good sleeper, that transition between alertness and slumber is ever-elusive, especially when you want it the most. When I began writing this book, my plan was to explore ever-more-efficient techniques to fall asleep quickly for short and effective naps. Napping soon became my entry point for understanding the transition between being awake and dreaming. Regardless of how it ended up, the number-one complaint I got from people when I told them I was writing this book was that they couldn't fall asleep when they wanted to.

Try to nap and you might instead shift around as your mind wanders aimlessly until the appointed time when you know you have to get up. Or maybe you have the opposite problem. You might fall asleep so fast and hard that you wake up feeling like you're underwater. Everything is sluggish, and you just can't shake the symptoms of sleep inertia —a symptom of not completing a full sleep cycle once you've already gotten underway.

As Americans accustomed to quick fixes—and, heck, maybe as humans in general—the ideal solution to this age-old problem should come in the form of a fast-acting capsule that you can take right before you want to head to sleep and that doesn't come with any grog-inducing side effects. The pharmaceutical industry has an entire arsenal of drowsy-making medications, from somnolescent antidepressants like trazodone to chemically addictive benzodiazepines to a new generation of sleep inducers like Ambien (zolpidem) and short-acting Sonata (zaleplon). Your body will absorb every single one of these chemicals in different ways, leading to a wide array of physiological outcomes that resemble natural sleep, but they also all come at a price. Sleep scientist Matthew Walker cautions against using sleep aids at all. In his view, while they can occasionally be useful in limited short-term circumstances, their tendency to foster dependence on a drug instead of the natural oscillations of the body ultimately make long-term use less promising. In some cases, they're even dangerous.

So I began to wonder: What if I could devise a pill that would send a person to sleep without any of the nasty pharmaceutical side effects? What if the pill wasn't a pharmaceutical at all? Every drug that targets sleep on the market shares a common quality: They operate by manipulating chemical mechanisms in the body that change brain chemistry or your overall hormonal balance. They are pills that cause sleep without any regard to anything that a person actually thinks or does. There's no conscious component—

you just take a pill and let your body's chemistry do the rest. The problem with that approach is that, as I mentioned in the prior chapter, sleep is fundamentally a conscious experience. Sleep isn't like hitting an off-switch in your mind; rather, it's a series of transitory states that come with their own physiological components and produce sensations that you actually experience (even if you don't remember them later).

At best, you might take a pill that's a ticket to the journey through various sleep stages. At worst, the sleep that comes in pill form is more akin to true unconsciousness—a tranquilized state more like a coma than a dream. So I wanted to know if I could change the logic of what a sleeping pill is in the first place. Instead of relying on pharmacological properties, I wanted to see if I could maximize the underlying principles behind the placebo effect to see if a person could cause their own sleepiness.

That might sound strange, so let me explain. Since at least the 1970s, some doctors have become fascinated with the idea of so-called active placebos. Active placebos don't initiate any important mechanical or biological activity of their own, but instead provide sensations inside a person's body that the placebo-taker can latch onto so that they think the pill is causing the desired outcome. In other words, active placebos create sensations where drugs like penicillin work whether or not you feel anything when you take them.

For example, the integrative physician Andrew Weil suggested in 1986 that the supposed medical benefits of marijuana might simply come down to a placebo response (a notion that was recently upheld by the psychedelic-friendly journalist Michael Pollan in his book *Your Mind on Plants*). He posited that the euphoric effects of cannabis help with pain, anxiety or depression simply because some people expect those effects. He noted that other people who have negative impressions of marijuana can instead have a negative ("nocebo") reaction that leaves them more anxious and in more pain, or can deepen their depression. Subsequent research has found many so-called active compounds in the cannabis plant, but Weil's assessment of marijuana remains prescient, and suggests that focusing too much on compounds and chemical pathways neglects the reality that sometimes change happens because of the things we feel, not the things that happen to us. Perhaps there is no "real medicine." Instead, our expectations are just as important as the chemical pathways that drugs work on. As far as patients are concerned, healing is healing.

Which is why most of the research on the effectiveness of the placebo effect also has to account for the role that medical rituals play in positive health outcomes. Even when a clinical trial does its utmost to factor out unintentional bias as it zeroes in on the objective healing power of whatever medicine it is testing, the patient receives their pills from a respected medical authority, usually in the ritual garments of science—a lab coat or medical scrubs. The set and setting of the intervention happens within the context of scientific medicine that we already associated with healing.

In the past, doctors believed that the placebo effect required tricking patients into believing they were taking a "real" medicine. This is why in clinical trials researchers give placebos to one group and active drugs to another without telling any of the subjects which group they are really in. More recent research out of Harvard by Herbert Benson showed that patients who were told that they were receiving a placebo for their irritable bowel syndrome still saw their conditions improve despite knowing the medication was inert. In other words, simply the act of taking a pill and thinking good thoughts about it fostered the desired outcome.

So back to my idea.

What would happen if I could find an active placebo that someone could take as they fell asleep at night that triggered mild sensations that kick in during stage 1 sleep—the transitory stage between sleeping and wakefulness? Just maybe, their mind would latch onto those novel sensations and associate them with the plunge into a dream state. Once the brain makes those sorts of associations between the active placebo and sleep, then maybe you could take the pill to trigger the underlying sleep conditioning. Thus, even though the sensations were essentially neutral of any pharmacological sleep property, when you took the pill, your brain would respond to the conditioning like a Pavlovian dog.

It's what psychologists call second-order conditioning, and it at least deserved a try.

So I reached out to volunteers to see if they were interested in trying out my napping experiment, narrowing my focus to a few supplements that I knew triggered mild sensations. I was most interested in a supplement called beta-alanine, which dilates the arteries in the extremities and causes mild tingling sensations. But I also included magnesium and 5-HTP (a precursor to melatonin that reportedly has some sleep-inducing sensations). I told all the volunteers to take the supplements for a week, just when they were about to go to bed each night, and then perform an eye-movement meditation called the Rosenberg Protocol that I hoped would accentuate a placebo response.

Sidebar: The Stanley Rosenberg Protocol for Sleeping

Lie down and very gently turn your head from side to side, and pay attention to how the movement feels. Then interlock your fingers behind your head with your arms open behind your ears and your elbows facing outward. Take a moment to feel your body in this position. Then, while keeping your head centered, move your eyeballs to the right as far as they can go—looking out over your elbow—and stay in that position for thirty seconds to a minute. If you feel a desire to yawn, swallow or sigh, bring your eyes back to the center. If you don't feel those sensations, don't worry, just return to the center after about a minute. Take another moment and then repeat the same movement on the other side. Move your eyes all the way to the left—looking over the left elbow—but with your head neutral for thirty seconds to a minute. Bring your eyes back to the center and then turn your head to the right slowly and then the left, taking care to notice any difference from how you felt doing the same motion at the beginning. Now let yourself drift off into sleep.

OVER THE COURSE of two months, I asked my thirty volunteers to first set a baseline of using the various interventions at night and to then use the protocol during afternoon naps. They all recorded their results in online forms, and I hoped that I'd get some interesting data. As I expected, just about everyone who actually took naps reported that they felt better after lying down for as little as just ten minutes, even if they didn't fall asleep—with 22% saying that a short nap was fully rejuvenating. But when I looked closer at the data, I saw a fascinating pattern: Every single person I'd assigned to beta-alanine dropped out of the program without even trying it.

This was a little surprising. While it's totally normal for people to drop out of self-administered studies, it was statistically odd for an entire group to disappear. So when I asked them what happened, they sheepishly told me that they all had Googled beta-alanine before taking it and found web pages that called it a stimulant. They were nervous about even trying a supposed stimulant before bed, and had decided that the study wasn't for them.

In other words, every single person who signed up for the study did what just about every sane person in the modern world does, and tries to reason for themselves what is about to happen when they take a new medicine. Unfortunately, this meant that the entire group started their conditioning by training a nocebo response instead of a potentially beneficial one. At the very least, this meant that I had a messaging problem. No matter how much I tried to set expectations up front, the reality is that people have their own ingrained ideas about what will help them fall asleep that likely would have affected how they felt about the tingling sensations from the drug if they had actually taken it. After three months of emails, surveys and data analysis, my active placebo idea was doomed to failure.

But there was a silver lining.

The Rosenberg protocol proved to be a hit all on its own. Even though I hadn't expected it to do much more than provide a short and easy ritual before going to bed, it turned out that everyone who tried it thought it helped them. After all, Rosenberg's method was a placebo in its own right. The easy head and eye movements paired with their intention to take a rest prepared the way for sleep much better than taking a pill.

The takeaway for me was that, even if an active placebo might, in theory, work to help someone fall asleep, a meditation and ritual practice would probably do the same thing, possibly even better. While I think that just about anyone can try out Dr. Rosenberg's technique, there are other sleep meditations that have been in use for hundreds, if not thousands, of years that are much more effective.

5

NOW START DREAMING

Since the quick fixes to our failure to catch sufficient shut-eye don't seem to be fixing things quickly, it's worthwhile to revisit the most tried-and-true advice on the market today. After all, the most popular books on sleep will spend hundreds of pages telling you why you should be getting sleep and offer a few tips on practices that might disturb sleeping patterns, but they don't have much to say at all about what it actually takes to fall asleep.

General guidelines go like this: Stick to a regular sleep schedule, prepare a safe and comfortable place to rest, don't eat food or drink alcohol right before you go to bed, limit screen time, have softer lighting that eliminates the bluer frequencies, and wind down your activity level at least an hour before your head hits the pillow. Researchers note that sleeping pills can help in the short term but warn that, over the long term, their addictive properties will make it harder to sleep overall.

A curious pattern links all these pieces of advice: They're not really about sleeping. They're about making a better bedroom and preparing your headspace for turning down.

It turns out that scientists are far more comfortable looking at so-called objective data—from electrical conductivity on the skin and the shape and temperature of a room to rapid eye movements and whatever else they can measure—than they are talking about the experience of sleep itself. The advice is good, of course: The more comfortable your room is, the more likely you are to have a nice sleep. But it's not really what you're looking for, is it? You want to know how to fall asleep, not where or when is the best time to do it. It turns out that this is the trickiest question of all, because we're not actually interested in sleep physiology; rather, we want to change our mental state from being awake to being asleep. It's a question of consciousness, not physiology.

The problem that most of us encounter happens when we've already made the intention to go to sleep. We close our eyes and get completely frustrated that the perfect state of non-doing just doesn't come. Like you, I've lain in bed for hours with my mind racing from one topic to another, wishing that it would all just turn off for a few minutes so that a soporific wave of bliss could carry me off into the sleep realm. I, like you, want a trick that expedites the process of unconsciousness. We both know that sleep is good and restorative, and we want to get there as quickly and efficiently as possible.

Of course, the times when I most need sleep are, paradoxically, also the times when it's hardest for me to achieve it. When I'm stressed out because of something at work, if I'm sick, or if someone said something wrong on the internet and I feel the (unproductive) need to correct the record, my mind can spin in endless circles and loops. When I'm at my most stressed out, I think I can even feel parts of my brain grow tired as the same synapses fire relentlessly over and over again. My waking mind chews and chews on a problem as I play out different scenarios in my mind about how other people will respond to my most clever retorts. Sometimes these midnight plans actually do result in things that I can take action on in the morning, especially if I write them down. But in general, the endless loops don't serve me in the same way that sleep could. And I know it.

Unfortunately, worrying about all the consequences of a bad night's sleep is a tried-and-true way to never actually fall asleep. There's a sort of negative voodoo that happens when a person pushes on the desire for sleep so forcefully that they drive themselves awake trying it.

In a perfect world, we would all be able to drift off the moment we felt tired and put our heads on a pillow. We've been practicing falling asleep and waking up since before we were born, but most of us have difficulty describing exactly what we do to fall asleep. How do we define that skill set?

The most basic form of mindfulness meditation that pretty much every yoga teacher and meditation guru teaches is also, arguably, the most powerful for sleep, because it's specifically designed to break thought loops. The yogic insight is a deceptively simple change of perspective. Instead of being your thoughts, you decide that you are instead the watcher of your thoughts.

Let me explain that in another way.

When you're stuck in a thought loop, one thought leads to another, then another, in an endless, vicious cycle. Perhaps it goes like this: "Oh, man, my bills sure are adding up, but I can't seem to earn enough to make ends meet. I should work harder. Or maybe I need to invest in something that will pay dividends. But I don't have enough money, and my bills sure are adding up." Your loop may sound different, but you get the idea. It's an endless cycle of thoughts where your brain keeps ending up in the same place it started.

Practicing mindfulness first means trying to make space between the velocity and motion of your thoughts and your ability to be aware that thoughts are happening. This might sound like some sort of split personality, but the more you try this, the more you will be aware that your own brain exists in multiple different perspectives at once. So instead of getting caught up in your own internal dialogue, you become an audience to that dialogue.

So how, exactly, do you do this? Trust me, it sounds trickier on paper than it is in practice. Let's try it. What is going through your mind right now as you're reading this page? Have any other thoughts intruded on your focused attention? Did you wonder what you're having for dinner, or if the laundry is done? Did an idea you read on the page here remind you of something you'd read in another book? Congratulations! You just observed your own thoughts! Now that you're aware that this is indeed possible, the next step is to actively interrupt the cycle.

Before I try any specific techniques, I start out with a clear goal. Or maybe you'd prefer to call it an "intention." Either way, once I've identified a thought loop, I tell myself that this thought loop does not serve me and that dreaming will offer me a fresh perspective. This is a way to give myself permission to fall asleep.

After that, I choose a specific set of sensations in my body to focus my attention on, but, critically, I don't attempt to define those sensations with words. I just try to experience them as raw, unfiltered feelings. Sometimes I will focus on the deep space that I can see behind the lids of my eyes or the tiny red and blue flecks that appear in my visual field out of nowhere. Sometimes I will try to move my mind forward into that space as if I were going on a journey. If I feel a sensation in my eyes or a soft rumble in the back of my head, I encourage the sensations and try to go toward them with my mind. I let my thought loops recede and don't attempt to form words out of them. It can take a little practice to become an expert thought breaker—but luckily, you have an entire sleep cycle to get good at it.

In yoga, there's a concept that I love called drishti—which I'll translate roughly as "the object of focus." If you're in a yoga pose that requires an element of balance, such as standing on one leg, you will find that looking at a single unmoving point on the floor will help you stay upright. This point of focus is a drishti.

When you're endlessly stuck in a thought loop, you've unconsciously allowed that thought loop to become your drishti. The more you focus on that drishti, the more stable it will become until the loop drowns out all other possible focal points. All you need to do to break the cycle is to change your drishti.

Believe it or not, the childhood solution of counting sheep is a pretty decent drishti. I know it sounds silly, but if you could visualize some fluffy white sheep jumping over a fence every time one of those cyclical thoughts attempts to crop up, you would have a pretty powerful practice to break thought loops. If counting sheep feels too childish, try counting breaths. Pay attention to the rising and falling of your chest as you breathe. Feel the air move in and out of your lungs. Notice how the air that enters your nostrils is cool and the air that exits is warm.

There are other specific breathing techniques that can help calm the body and the mind. One you could try is called box breathing. In this pattern, you breathe in for four seconds and then hold at the top of your inhale for four seconds. Then exhale for four seconds and hold your breath at the bottom of the exhale for four seconds. Repeat three or four times and then see if you are relaxed enough to drift off into sleep.

Finally, if you absolutely cannot shut off the monkey mind of incessant thoughts and words, you could try a mantra instead. A mantra is simply a word that often has no inherent meaning that you repeat to yourself instead of the random chatter in your head. In the yogic tradition, the syllable "Om" has been used for millennia. Another option is to make the word "dream" your drishti instead. Or literally any other word that you happen to like. Focus on it every time a different set of words enters your mind.

Drishtis are a sleeper's superpower. The reason they work is because they are a type of non-working. They don't build into anything more than what they already are. You simply use them to distract your waking mind from repetitive chatter so that the natural process of sleep can find space to take over your consciousness. They work just as well when I'm trying to fall asleep at night as when I wake up at three or four in the morning, or when I'm lying down for a short ten- or fifteen-minute nap.

Then again, there is no reason that you need to draw on the wisdom of ancient India to do something that should come naturally to you. Another technique that's equally compelling and effective is to think of sleep as a journey instead of a destination. In the moments after you close your eyes, you can start to tell yourself a story. Remember how effective a bedtime story was for putting you to sleep when you were a kid? Or maybe you have seen it with your own children. Bedtime stories are an ancient technique to prime a person for dreaming. And there is no reason you can't tell yourself your own bedtime story. Try to visualize yourself in a place that is familiar and comforting—like a placid field somewhere near the home where you grew up. Tell yourself a story about that location and see where it leads you. Perhaps you see an animal. Follow the animal to see where it goes. Look up at a cloud and observe it transforming into whatever your unconscious mind wants. You will be surprised at how quickly telling yourself a story can turn into a hypnagogic dream. With any luck, you'll be asleep before you know it.

Regardless, if these specific pieces of advice ultimately work for you, remember that you've trained yourself for sleeping your entire life, and that somehow you've achieved the perfect slumber in the past. You're already an expert at moving between different states of consciousness, and whether you remember your dreams or not, you've already drunk their sweet, restorative nectar. Your sensations in the middle of the day that warn you of your flagging energies are part of the process that will gently guide you into the perfect sleep. Your only job is to pay attention to the messages that your body is already sending you and to let it happen in the same way that you've let it happen in the past.

Remember, it's totally normal for a person to go through times when sleep comes easier and times when it comes harder. Sometimes the body needs more restorative time, while at others it's raring to go and tackle the tasks at hand. It's your responsibility to listen to yourself and decide what feels right for you right now and what will make you feel better in the long run. You just need to believe in yourself and let the obstacles that don't serve you melt away into sweet unconsciousness. Your dreams will take care of the rest.

6

THE POWER OF RITUAL

The anthropologist Matthew Spellberg spent years studying the dreams of a remote tribe called the Onge who lived on an island in the Bay of Bengal. After a typical night, members of the tribe would gather together and talk about their dreams. For them, wrote Spellberg, "dreams are understood to be sites of action; not texts but places, not a coded language, but a part of reality." Every night the Onge would visit a dream version of their island and then tell each other about the events that happened there when they woke up. That constant reinforcement made the events and symbols of the dream island a parallel reality that was as real as their own home. More important, they cultivated the dream space together.

It's hard for someone accustomed to the constant stimulation of the modern world to truly accept the existence of a shared dreamland. We all agree that the social lives that we piece together on Facebook or in video game worlds occupy a shared slice of reality, but we balk at the idea that dreams might be similar. Maybe that's because we feel like we remember the sequence of events that happen as our awareness shifts when we use a computer or log on to a website. But take a second and think about whether that's really true. Do you really see what's happening on the periphery of your vision when you're sitting at a computer, or have you used the strange alchemy of consciousness to limit your experience to just what is happening on screen? If someone calls out your name from across your house, is it jarring to shift between the real and virtual worlds? The shifting of awareness that we're so used to when using technology is similar to what happens during the natural process of dreaming. For their part, the Onge would have no trouble recognizing dreams as the first virtual world.

Of course, one major difference between dreaming and looking at a computer screen is that you can usually remember what happened on your computer screen a little more clearly than last night's dream—at least in the short term. But things get a lot blurrier if I asked you to remember what you did on your iPhone three days ago. Sure, you might have scrolled through a hundred images on a social media app, but do you actually remember any of it clearly? Even when we don't remember what happens in the dream world, the act of dreaming is the foundation of our emotional universe and future behavior. As the neuroscientist Roland Benoit once said, "Our memory is not made for the past, but for the future." Moreover, who we are in our waking life changes according to our context and environment. You are a different person on a battlefield than you are in a flotation tank. Your most heartfelt desires morph between goals, depending on whether you are courting a romantic partner, meditating on the nature of enlightenment, receiving your college diploma or interviewing for a new job. All of those contexts define the boundaries of who you are in any given moment.

Dreams are no different. They're a place where it's possible to seamlessly flit between an almost infinitely diverse set of contexts and thus experience a new version of yourself in every single one. This is also probably why it's so hard to describe what happened in dream life when you wake up, and why hypnagogia inspired the surrealist movement.

When we're awake, time seems to move forward in one direction, and the transitions between who we are from one moment to the next retain a measure of continuity. In dreams, it's entirely possible, or even ordinary, to perceive yourself in a different body—with a smaller or taller frame, a different gender, or the ability to fly—while at the same time moving between different time frames. One second you might be in your childhood bedroom. In the next you could be back in the cafeteria at college, and a few steps later, behind the wheel of your first car or tasting the most exquisite slice of cheesecake you ever could have hoped for.

The most startling insight between all the changes in context and content, at least for me, is how ordinary the transitions feel. In dream space, the dreamer simply accepts the premise of whatever environment and activities the dream presents. As the mind wanders in this other environment, the body in the physical world remains paralyzed, and none of the signals from the motor cortices make their way to the peripheral nervous system.

Normally, we like to think of consciousness as an extension of physical existence. But what does it mean that an incredibly essential and regular biological process in our minds literally moves our bodies out of their physical context and into environments that our minds conjure up out of memory and stored experiences?

Dreams make us reconsider the location of the mind

altogether. Dreams feel real because, as far as the wiring in your brain is concerned, dreams are real. Remember: The brain does not sense the world directly. The brain has no sensory nerves of its own. If you happen to be awake during brain surgery, you can't feel the doctor's finger poking your brain matter. Instead, the brain (and your consciousness) only experiences its surroundings through the signals that come in through the peripheral nervous system. At its most basic level, consciousness is a simulation that your brain compiles of the world at large. This holds true for what you feel in your body itself. Should someone hit your toe with a hammer, the pain you feel happens in the brain, not the toe.

But wait—it gets even weirder than that. It takes a certain amount of time for the peripheral nervous system to transmit sounds, sights, smells and sensations from the outside world into your gray matter for processing, which means that your consciousness doesn't experience its surroundings in real time. It takes, on average, about a fifth of a second for all that information to arrive in your brain.

A fifth of a second isn't a huge time gap, but it can sometimes be enough to make the difference between life and death. To make up for the lag, the brain does a remarkable thing: It speeds up time. Since the brain is already running a simulation of the world, it also has the ability to make predictions on how events outside the body will unfold. In order to act in as close to real time as possible, the brain over-clocks its simulation based on its best guess of what will happen so that it can run a fifth of a second fast. Let me give an example. Let's say you're playing catch with a friend. When they toss the ball in your direction, your eyes detect the light reflecting off the ball, and that data gets translated into chemical and electrical signals that make their way to the brain, which then puts together a response to move your hand to just the right place at the right time to catch the ball in the air. But if your hand arrives at that spot a fifth of a second late, you'll miss it altogether. To correct for this, the brain anticipates where the ball will be and seamlessly speeds up time so that you experience a virtual simulacrum of real time and can catch it.

This is why Adam Haar Horowitz says that conscious awareness during your waking life is just another form of dreaming. Sure, the things you see and feel during the day happen outside the body, but your experience of the world is a simulation in the brain. From this perspective, a consciousness provocateur could argue that the purest expression of the self isn't what you're feeling and sensing right now. Pure consciousness only happens when the brain experiences itself in dreams.

If that sounds weird, don't fall into an existential crisis just yet. We're only getting started. Over the centuries, humans bridged the expanse between the dream world and the ordinary one through rituals. The Onge's collective dreaming is just one example. Similar practices occur with aboriginal groups in Australia and North America. Meanwhile, the ancient Greeks and Egyptians constructed elaborate sleep temples as gateways to predict the future and heal the sick. Thousands of pilgrims traveled to the temple (called an incubado) of the doctor-deity Asclepius in Greece for a chance to cure their wounds by napping at his feet. They would report their dreams to a priest who would use the symbols of the dream world to evoke real healing in their bodies. Hippocrates, the father of Western medicine, trained at such a temple on the Greek island of Kos. The Rod of Asclepius—a scepter with two snakes below bird wings—remains the professional symbol of most modern medical associations around the world. And, as mentioned in earlier chapters, since various sleep stages correlate with repair and regeneration in the body, it stands to reason that the Greeks were on to something when they saw the roots of healing in their god of medicine and sleep.

Most physicians today would argue that whatever healing happened at those Asclepion temples was nothing but the placebo effect in action. For the purposes of talking about sleep, remember that ritual is a bridge into the dream world. Rituals do not need to be elaborate to move people between subjective and objective experiences. They just need to be part of a routine that sets up expectations. Sleep researchers have shown time and again that the more regimented and ritualized a person's activities are before bed, the more soundly and quickly they will sleep—and, should they want it, the better ability they have to incubate dreams and influence what occurs in dream space.

Conversely, if we go to sleep without rituals and instead infuse the time before bedtime with activities and thoughts that push us away from sleeping—whether it's surfing the internet on our phones, playing video games, drinking or engaging in vibrant debate with our bedmates—it becomes all the more difficult to actually make the transition into dream space.

In neuroscience, there's a distinction between perception—the signals that come in from the outside world to the brain—and conception, the way that the mind creates models and beliefs about the world from within itself. Perception is outside in, while conception is inside out. These two processes happen almost continuously and seamlessly during waking hours, in an endless conversation that helps form our experience of reality. Conversely, the dream world exists almost entirely in conception, as it chews through the material that we perceived in the past.

Thinking about rituals in the context of these neurocognitive processes suggests that rituals alter the environment and prime the mind for conceptual thinking—which, in turn, alters the physiological responses in the body. This is why sleeping rituals guide us easily into the dream world.

In ancient times, in places like Delphi, North Africa and Tibet, people with important questions that didn't have obvious solutions went to oracles looking for answers. After the appropriate offerings were made and the importance of the question made apparent, these oracles typically consulted the spiritual realm in their dreams. (Other shamanic practices used hallucinogens, breathwork and other trance-inducing methods for similar results.) Barring any supernatural explanations for what happened during the ritual dream time, a neuroscientist or anthropologist would look at the ritual through a psycho-biological lens. In rituals, the human brain reduces all of the data from the day before into emotional gists during the dream state. This emotional forge combines the disparate sensory information that it collected in short-term memory during the day, as well as other sensory data that a person might never have put words to. With careful preparation, the sleeping oracle would have more access to remembering the content of the dream, but, more important, would emerge with emotional imagery that he or she could relay to the person asking the question.

Oracular proclamations may not have been perfectly accurate in terms of predicting future events; however, the mere fact that they survived to form the backbone of cultural institutions is evidence enough of their utility. The neurological explanation of oracles suggests that their prognostications were actually based on data compiled by the brain in novel ways that ultimately came up with predictions and courses of action that were more likely to be beneficial than harmful. In other words, assuming the brain works for the benefit of the organism it inhabits, oracular dreaming operates with better odds than pure chance.

This isn't really much different than other predictive tools we use in the modern world. Most artificial intelligence applications, for instance, search for signals in immense pools of data, based on whatever written inquiry or prompt the user suggests. The AI makes abstract links between data until its recommendations home in on a single output. In most cases, AI recommendations are not able to show their work for how they came up with a particular conclusion, but the user trusts the AI insofar as they trust the process. It is somewhat amusing to think of AI as perhaps just a slight innovation on truly ancient techniques.

But no matter how many AI programs use dream metaphors in their brand names, (Midjourney and Dream AI are just two examples), this is a digression, because every human already has the biological ability to access intuitive thinking.

So, how do we create sleep rituals that help us alter dream states?

There are a variety of practices to draw on throughout the ethnographic record, but rituals from foreign cultures do not necessarily translate to the modern world or, for that matter, your own bedroom. To do this, I would recommend creating your own rituals that speak to you uniquely. The only thing that matters when creating your own ritual is that you believe that it will work and that the ritual is meaningful. This can sound a little strange to someone used to mechanistic explanations. It can even sound a little pseudoscientific. Yet study after study has shown that the people who are most pliant and suggestible are the ones who are also most likely to benefit from a robust placebo response.

As mentioned in an earlier chapter, the targeted dream insertion that Adam Haar Horowitz dreamed up at MIT's media lab offers one such intervention that might speak particularly well to someone brought up on Western thought. Dormio, the glove-shaped device he created to detect the first stages of sleep, along with audio prompts from a typical smartphone, was reliably able to trigger specific dream content in 97% of the subjects. The concept behind it was deceptively simple.

Just like ancient Greek oracles, the prospective dreamer begins by setting an intention for the content of their dreams. Haar Horowitz asked his subjects to dream of trees by imagining before they fell asleep lush leaves, deep forest groves and anything else that might elicit a tree. The more preparation and sensory input, the more effective the dream would be.

Hypothetically, this might mean that if you wanted to dream about a dead relative rather than trees, you might want to listen to recordings or watch home movies of that person. You might even smell clothing or a cologne they once wore to prime the brain's short-term memory storage. Alternately, if you want to learn a foreign language, listening to or reading that language before sleep might also help prepare the way. The same goes for any new skill you want to learn.

Whatever the case, when using the Dormio, the subject would record an audio prompt into their smartphone such as "Dream about your mother" or "Quiero soñar en español" ("I want to dream in Spanish"). The person then falls asleep to the intention of whatever vision they are hoping to incubate. About seven minutes after the Dormio detects changes in skin conductivity and heart rate indicative of stage 1 sleep, the device triggers the smartphone to play the audio message. A few minutes later, another message wakes the sleeper up and turns on the audio recorder so they can recount what they were dreaming about before falling back asleep.

Haar Horowitz's experiments primarily targeted hypnagogic sleep during the first phase, but it appears equally possible to use the same process at any sleep stage. Indeed, it doesn't seem necessary to wear Haar Horowitz's elaborate glove to bed in order to trigger a dream. Just about every wearable sleep tracker on the market right now can detect when a person enters various sleep stages based on heart rate variation alone.

Since we know that different sleep stages correlate with different kinds of memory and motor processing, it doesn't seem like a huge lift to craft specific prompts for optimal results. For instance, an athlete looking to improve their performance in sports could trigger a dream around a particular athletic feat they're working on during stage 2 sleep if the skill is something they're familiar with, or during REM if it's an entirely new set of movements that need more attention. A person suffering from PTSD might want to relive the difficult memories during the first or second REM stage, while a person with an autoimmune illness might benefit from a sleep prompt during the deepest stages of sleep.

Unfortunately, there are some limitations with this invention out of MIT. Most notably, as of the time I am writing this book, there's no device on the market that actively attempts to incubate dreams this way though audio prompts and recordings. My own attempt to build a Dormio through the open-source instructions that Haar Horowitz posted online ended in a hilariously ineffective wild goose chase. I hired someone to build the device for me, only to discover a few thousand dollars later that the software was so out of date that there was no hope of it connecting properly. I reached out to several sleep-tracking companies (notably, the science team at Oura Ring and the people who develop content for the Apple Watch) that were intrigued by the idea but beset by the typical institutional inertia that makes new features slow to implement. Nonetheless, I fully expect that some entrepreneurial tech investor will soon bring a device to market for easy-access digital sleep rituals.

The opportunities to self-experiment with learning during sleep seem almost endless, and, at the same time, have few downsides. While most clinicians sing the praises of sleep for recovery, resilience, immune function and cognition, there are not a lot of direct therapeutic interventions that target dreams specifically. One caveat, of course, is the renewed interest in psychedelic medicine for mental health problems, which seems to be on the fast track for FDA approval as I write this chapter. While psychedelics are different from dream states insofar as they originate with a chemical substance, the actual content might not be much different in practice from what happens during dream states.

Dreams are still an untapped space for changing the way a person relates to the waking world. Although clinical research on dream incubation is still in its infancy overall, the side effects of failing to incubate a particular dream or incorrectly incubating a dream or any other response that might happen to an audio prompt played over a smartphone pale in comparison to the side effects of almost any bio-active pharmaceutical or even supplement.

That said, there is another ritual that I believe might be even more effective than the Dormio and its kin that you can try right now.

7

YOGA NIDRA

Falling asleep is a little like letting a camera lens go out of focus: Sleep stages blur at their edges between alertness and oceanic experience. For the most part, we have only a limited ability to recall how each stage feels. Consciousness slips away to unconsciousness as naturally as water pouring out of a pitcher. What scientists know about different sleep stages relies mostly on the output of various brain scanners and what people can recollect afterward as their short-term memory quickly fades. Neither data set is entirely satisfying.

However, there is at least one meditation technique that seems to allow a person to stay aware of sleep stages all the way down to a point where most scientists don't believe it's even possible to have awareness at all. Brain scans of practitioners have recorded delta waves—the longest and most languorously infrequent electrical impulses we can detect in the brain. What's more, it doesn't need to take a long time: People can seemingly drop down and come out of the meditations without the adverse effects of sleep inertia.

The practice is called Yoga Nidra, which roughly translates from Sanskrit as "sleep meditation." The yoga that most people are familiar with involves a series of postures that limber up the limbs, which usually ends in a pose called savasana (also known as "corpse pose"), where you lie on your back and let all the tension drift out of your body.

At a basic level, Yoga Nidra is a mental, spiritual and physical inquiry into that one position—an exploration and journey into the deepest possible states of relaxation while maintaining a certain level of conscious awareness at the same time. To learn about it, I reached out to the Stanford neuroscientist Andrew Huberman, who in turn suggested I talk to Kamini Desai, author of the book *Yoga Nidra: The Art of Transformational Sleep*. In her book, Desai is clear that she sees these techniques as a sort of nap on steroids, writing that "using Yoga Nidra as a nap is like using a jet plane to drive to the grocery store." This got my attention, partially because the way I understand the benefits of sleep is that they're powerful precisely because they happen during unconsciousness. You're not supposed to remember your dreams. Dreams create gists of memories and forge emotions. But the benefits of sleep go beyond just what happens in dreams. Sleep restores the immune system, clears waste out of the brain, fixes worn muscles and a thousand more things. It appears that Yoga Nidra instigates an intermediary state of consciousness distinct from anything else I've talked about in this book. PET scans and EEGs of practitioners show restorative brain wave patterns that occur simultaneously with conscious recollection.

In one study, Danish neuroscientist Troels Kjaer scanned the brains of nine Yoga Nidra practitioners who all had more than five years of experience with sleep meditation. By identifying different brain areas that lit up during meditation (and inside the bulky and sometimes noisy MRI machines), he showed that the perception areas of his subjects were much more active than would be expected, while the areas of the brain that controlled muscle groups and movement were quite low. This demonstrated the paradoxical findings that Yoga Nidra made people incredibly aware of themselves while simultaneously allowing them to be so relaxed that they were almost indistinguishable from sleeping subjects.

A second study Kjaer ran found that Yoga Nidra practice increased dopamine release by 65%—which indicated that people practicing Yoga Nidra were, at a chemical level, seeking a reward. While the nature of that reward is unknown, it could be evidence that the practitioners' intention to restore their spirits had a physical output.

Whatever underlying mechanisms they might reveal, brain scans offer only a limited window into how a person's quality of life changes once they start a new practice. Fortunately, qualitative assessments are easier to come by. Richard Miller is a psychiatrist, as well as the founder of the International Association of Yoga Therapists and the Integrative Restoration Institute, which teaches Yoga Nidra practices to combat veterans, people struggling with addiction, and as a general therapeutic intervention called iRest. Research he conducted at Walter Reed with the Department of Veterans Affairs showed that these practices could decrease chronic pain by as much as 40%, as well as radically improve the symptoms of PTSD.

Desai's book (and others on Yoga Nidra) draw on ancient Hindu traditions that go back at least 5,000 years. Even so, there is some disagreement as to whether what people practice today under the Yoga Nidra moniker is the same thing that ancient Himalayan sages recorded in the Rig Veda, whether they're more recent inventions from Swami

Satyananda as late as 1940, or if they arose somewhere between those two extreme dates. Perhaps this confusion makes sense. Sleeping and napping are as natural to human existence as breathing and eating. There's no question that people have tinkered with sleep since the beginning of human consciousness and will continue to do so as long as we are around.

Regardless of its origin, I have to admit that Yoga Nidra is transformational.

It just so happened that I spoke with Desai one day after I came up positive for COVID (don't worry, it was a Zoom call). It was a mild case, but my entire body ached, I had a fever of 101, and I hadn't had a good night's sleep in over a week. The questions I'd jotted down before the call centered around the apparent contradiction that Yoga Nidra presents any would-be dreamer: How can someone be both asleep and aware at the same time? Since dreaming is a place where free associations happen in the brain, while awareness focuses attention on discrete endpoints, the very idea of entering deep stages of sleep while remaining alert seemed unlikely. I was also confused by the exact technique to achieve these unusual states.

She explained that when a person first learns to control their sleep states, they usually begin by listening to an audio recording or to an instructor while they are in a comfortable position in bed. The teacher then leads the sleeper to increasingly deep sleep stages while asking them to remain focused only on their voice. This sounded like a form of hypnosis to me. Desai acknowledged the similarities but said that Yoga Nidra was also a practice that someone could learn to fall into without guidance once they had a little practice.

"It's just like sleeping," she said. "It's hard to explain to someone else what exactly you do to fall asleep at night. I like to think of it as you finding a way to dive deeply into non-doing altogether." That is, of course, the fundamental conundrum of talking about sleep techniques in general. How do you create a set of instructions around falling unconscious? At some point, you just sort of do it.

Which is exactly what she recommended I do once I got off the phone with her.

Half an hour later, I crawled under my covers with a pillow under my knees, with Desai's now-familiar voice softly speaking instructions out of my iPhone and a playlist called "Healing Shift." The meditation started out with some light stretching and then settling down into light breathwork, and then a body-scan meditation where I paid attention to specific body parts sequentially from my toes to my head. These common mindfulness techniques moved on to me focusing my awareness on various internal organs and then repeating affirmations to myself that the things I was doing in bed were unlocking the innate healing powers of my body. I played a delicate balance in my mind—trying to pay just enough attention to Desai's instructions that I understood them, but also letting myself automatically follow her commands.

Much to my surprise, it worked. I found myself in a deep state of relaxation where I was completely aware of her voice but not focused on anything other than my body. The practice was alien at the same time that it felt familiar, no doubt because I visit similar places every time I go to sleep at night.

After I'd completed a review of every part of my body, Desai's voice issued a sort of hypnotic command: "Know that this practice has initiated a profound healing shift in my body." The statement was an invocation and assumption that it's possible to communicate directly to the immune system and at the cellular level during deep rest. On one level, this sounds profoundly unscientific. Yet when the thought later occurred to me, I remembered that Haar Horowitz once said to me, "Sleep might be the ultimate source of the healing power of the placebo effect." Sleep scientists have long noted the rejuvenating power of sleep for the immune system and how healing happens at a cellular level during deep stages of sleep. Since the placebo effect operates, at least in part, by the power of suggestion, maybe Desai's mere act of suggestion that healing can happen during deep stages of sleep could nudge forward a positive immune response. These are questions that a world-class sleep lab could probably suss out, though I don't know of any studies to date that have attempted it.

That said, once I got up from 45 minutes of Desai's "Healing Shift" recording, I felt significantly more energized than I had over the course of the previous few days, when the symptoms of COVID made me feel lethargic, shivering, sensitive to touch and unable to exercise. Desai's meditation marked the turning point in my illness. That night, I slept a full eight hours, and I was well on the road to recovery by the next morning. That's certainly not enough data to say that Yoga Nidra cured my case of COVID, but it is a data point to at least consider. It certainly didn't hurt.

On that first encounter, I was able to dive deep into sleep stages that I was both familiar with and yet have never remembered specifically being in before. Clark Strand wrote about similar intermediate states of consciousness in his book *Waking Up to the Dark*. Lying awake at night during the so-called hour of the wolf led to different sorts of thoughts than what I tend to think about during my morning routine. The easiest way I can describe it is to compare it to when I wake up too early from a deep sleep and feel like I'm somehow underwater. My eyes are open, but everything feels sluggish and disorienting. Yet instead of feeling an urge to get moving, I'm completely alert and totally happy with being underwater. Yoga Nidra offers an opportunity to work inside your own mind, deep in the underpinnings of the subconscious itself. Maybe Yoga Nidra is a sort of deep state of suggestibility similar to hypnosis—with the exception that, with a little practice, you don't actually need a guide to take you into the state.

In the course of researching this chapter, I ended up talking to several Yoga Nidra teachers and researchers. Every conversation danced around the apparent contradiction of trying to learn a skill that is ultimately based on doing absolutely nothing at all. Understanding Yoga Nidra is like trying to define a doughnut by only talking about the hole. John Vosler is a Los Angeles-based wellness instructor, yogi and meditator who leads workshops all over the world and teaches Yoga Nidra as a method to reduce stress and regulate the nervous system. "But that's what Yoga Nidra does," he tells me. "Not what Yoga Nidra is." To understand the practice itself, he likes to posit big questions about the nature of consciousness. Being aware of sleep isn't a contradiction if you also realize that we have no definition of what it really means to be awake. "Yoga Nidra is sort of like going home. It's about waking up to who you really are." Like many things, it's something that you have to feel if you really want to understand it.

I asked Vosler how he performs Yoga Nidra when he's not using a guided meditation on tape. He closed his eyes and then considered the question. He began by describing a recent flight to New Delhi where he used the practice to make the trip more bearable. "I start by considering the environment I'm in." He visualized the cramped quarters of the fuselage, the passengers, the hum of the aircraft engines and then his own position in the aircraft—taking it all in his mind without passing judgment. Then he shifted his focus to his own body—the pressure on his back, the position of his legs and arms—before diving down to become aware of the spaciousness inside himself, his organs and the pulsating of his heart. "I think about my breath rising and falling. I follow it to the top of the inhale and then down through the exhale until I fall into a silent space of consciousness itself. I lose awareness." He told himself to float into this state of intense non-doing, and also reminded himself that he should come out of the meditation just slightly before he landed in India. It was sleep and it was non-sleep at the same time. "There is this part of my brain that reminds me when I need to come back and wake up."

Consciousness is an inherently slippery concept. Most of us experience the world from the perspective of the ego that pilots our bodies. That's one way to look at it, but we are also our connections to other people. Every thought that I've ever had originated in interactions with the environment and what I've learned from hearing other people's thoughts through communication. Nothing is ever truly original, but merely an iteration of something that came before it. In this way, even the things that I experience in the world have origins outside me. Whether you like it or not, reading these words means that I've, to some extent, entangled you with my own consciousness. Our synapses and neurons are wiring together through the medium of text on paper. This isn't magic. You alter other people's brain chemistry every time you interact with them. At a societal level, we are all just synapses of a superorganism of life itself on the planet. We are all part of a superconsciousness.

According to Vosler, Yoga Nidra takes this idea of a greater whole of interconnected thought and instead turns the lens inside the body. As you descend into the layers of consciousness inside your own body, the actions of your organs and the way cells sense their environment ultimately help your whole body hum along. Vosler says that Yoga Nidra lets you wake up to the reality of all the different layers of consciousness that make us who we are.

This is why Desai and Vosler's guided meditations start by focusing a person's attention on things in the environment and then draw the attention to ever deeper parts inside the body, starting with the breath and ultimately into the mind itself. In reality, Yoga Nidra isn't really about sleep at all. The sensations and brain states involved in becoming conscious in sleep inevitably lead to more access to awareness when you're awake, too.

Miller, the clinical psychologist who founded the related therapy program called iRest, discovered that his services were not only in demand from soldiers returning from the battlefield, but also people on the front lines. "I couldn't believe I was sitting in Walter Reed talking to a room full of generals and colonels about the writings of Patanjali and the Upanishads," he told me.

He told the assembly of officers that soldiers who train under him and become adept at Yoga Nidra will come out in one of two ways: "Some will realize that the people they're fighting are their sisters, brothers, aunts and uncles, and they put down their weapons or shoot over the heads of the enemies because they won't be able to do the job anymore. Others will continue to fight not out of anger, hatred or retribution, but because they have a job to do. And when the fighting is over, they'll put their weapons down and help the people that they had just been fighting."

Miller worried that the benefits he could offer weren't what the brass needed to train an effective fighting force. But when he looked across the room, all he saw was relief. "That was exactly what they were looking for. The Army wants to weed out the people who can't do their job; at the same time, they want the actual fighters to do it from the right perspective." Now he lectures to all branches of the Army and special forces. "The SEALS love it," he told me.

In the ordinary world, we focus on specific tasks because we want to accomplish certain well-defined objectives. A true practitioner of Yoga Nidra realizes that there's nothing that actually needs to be accomplished. Instead, all a person needs to do is welcome what is already here. "You don't need fixing. You just need to remember what you already know," said Vosler. This may be the only way to truly express what Yoga Nidra is, other than a supercharged nap. Instead of grinding away at an objective with the power of focus and grit, Yoga Nidra is a way to access the flow state, where everything happens easily, exactly when you need it to. It's a way to come home to your truest self. It's a way to experience the deepest parts of consciousness by simply trusting that your body already knows how to get there, just like it already knows how to fall asleep.

To end this book, I'd like to offer you a chance to try Yoga Nidra yourself. After all, reading is just another thing to do. Why not try sleeping instead?

8

IN DEFENSE OF DREAMING

How do you end a book whose entire purpose is to make you realize that everything that we have ever known about the world is, at some level, a dream? Whether it's because your consciousness clicks forward a fifth of a second behind reality, whether your visions of bodily healing take the form of ancient Greek rituals, or you imagine and cultivate an entire island in sleep with your fellow Onge tribesmen, all consciousness at some point connects to dreams. The most important thing to remember about your dream journey is that what we do in our sleep reflects what we do when we're awake. The constant feedback loop ultimately defines who we are. And this is particularly relevant to how we interact with the obligations of the waking world.

I'm not a sleep scientist. I'm not a revolutionary. And my guess is that you aren't those things, either. Instead, we constitute the other 99% of the world that is overworked, over-lit and torn between the obligations that we feel to be productive members of society and the messages that our body sends us every single day to prioritize rest. If there is only one thing that I hope you get out of reading this book, it's this: You have permission to dream. Putting your head down for a short nap or an uninterrupted evening of rest is natural, important, and should make its way to the top of your priority list.

Yes, sometimes it's inconvenient to set aside time from your schedule for rest. Sometimes, no matter how tired you might feel, you remember that the last time you slept, you didn't get the recovery time that you actually wanted. Maybe you still feel that prioritizing rest is somehow inherently lazy. Even while I've been writing this book, similar thoughts have crossed my own mind. Even while expounding on sleep's benefits by typing letters into a glowing computer screen, I've also put off my own desire to sleep in order to get some specific task done. I get it. We're human. All of us are beset by our unique concoction of desires, obligations and patterns. The beauty of occasionally dipping your toe into non-doing is that you don't always have to get on with the getting on of daily life. You can often just walk away and take a few minutes or hours to shut your eyes.

Once you start giving yourself permission to take those small breaks, you'll soon discover that the time you spend actively doing things will get more productive. You'll have more energy. You'll be healthier, and you will probably even look a little younger. You have permission to dream, because no one needs to give you permission to dream. You are a dreamer.

After you finish this page, I have one suggestion that might help a little bit with motivation. Put this book in a place you will come across during the day. Maybe stow it in a prominent place on a bookshelf near your desk, in the kitchen or some other conspicuous location so that your eyes might fall on its cover or spine when your energy levels begin to flag at some point in the day. I will consider my job a success if, at some point in the next few weeks, you see it and realize that right now is as good a time as any to remember that you're already dreaming.

9

WEDGE YOGA NIDRA SCRIPT

One of the best ways to practice Yoga Nidra is to develop your own practice. The following script was adapted from Kamini Desai's book *Yoga Nidra: The Art of Transformative Sleep*. I tested this script several times at public workshops. It is designed to be read to a group of nappers. If you would like to practice alone, the full audio version is available for free at scottcarney.com/dreaming.

Context

> *Close your eyes and quiet your mind.*
> *Be still.*
> *Relax. Trust. Let go.*
> *Breathe in fully and exhale with a deep sigh.*
> *Visualize the area around you right now—the people and objects in your immediate vicinity.*
> *Now, imagine you are looking down at where you are from high in the air like a bird.*

*Going higher still and seeing the Earth itself and
    knowing that you are there.*
*Know that the Earth is just one planet in a solar
    system in a universe of stars.*
*Give yourself over fully to the power of that
    universe.*
*Because in a very real way, you ARE that
    universe.*
*Again, breathe in fully and exhale with a deep
    sigh. Let go even more with each passing
    breath.*
Feel a deep sense of contentment enter your heart.

First Body Scan

*Now we will move our attention to the body, for
    the body is a universe unto itself.*
*We will move our attention into that body to
    release any and all tension as you give your-
    self over to the sound of my voice.*

Upper Body

*As you inhale, make fists and induce stiffness and
    tension throughout your shoulders, arms and
    hands.*
Tighten. Tighten. Even more.
Now let go completely. Relax.
On your next exhalation, relax even more.
    Let go.

Second Body Scan. Bottom to Top.

Put your arms by your side.
Now, as you inhale, deliberately induce stiffness
    and tension in your hips, legs and feet.
Tighten. Tighten. Hold. And let go.
Let go and relax completely. Relax.
Observe the flood of energy in your legs.

FULL BODY STRETCH

Extend your arms over your head.
As you inhale, stretch out as long as possible. Feel
    your entire body elongate.
Stretch...stretch...stretch...even more.
Now let go completely. Relax.
On your next exhale, relax even more.
Observe and feel the energy extending to all the
    muscles, nerves and cells of your entire body.
Know there is an entire universe inside of you.
Release and hold. Anywhere.
Pause.

THE HUM of the Universe

As you descend further into the conscious aware-
    ness of your body and consciousness, we will
    allow the sounds of our throat to invoke the

hum of the universe. You will follow my lead as you make a sound in your closed mouth. Feel how you can move the vibrations around your throat, your head and even your entire body.
(Allow up to seven breaths)
Now stop and be still.
Bring your total undivided attention to your whole body.
Those stimulating vibrations continue to extend through your whole body, just as vibrations continue throughout the universe.
Observe the energy field expand and grow. Extending everywhere and filling up every cell and nerve inside of you. Know that inside your body are unfathomable depths as small as the universe is great.
Now bring your attention to your eyebrow center.
(Instructor can touch the forehead of practitioner)
And empty your mind into a flood of energy.
Drop into complete silence and deep stillness.

Universal Instructions

As we enter this next phase of Yoga Nidra, you will remain as motionless as possible. If you need to make an adjustment, do it now, returning to a sense of inner stillness as soon as you are able.
Resolve to stay awake, staying in touch with the sound of my voice.

*Allow your entire body to respond directly and
non-mentally to my words.*
*Allow any disturbances, internal or external, to
draw you more deeply within.*
*Now shift from thinking and doing simply to
feeling and being. (pause)*
Do absolutely nothing from now on. (pause)
*Drop into the deepest state of tranquility, stillness
and peace in the third eye.*
Now your consciousness is in direct communication with the energy of your body.
*Now your consciousness is undifferentiated from
the universe itself. At all levels.*
There is nothing to do.

COMPLETE YOGIC BREATH

Now follow my guidance as we begin the relaxation breath.
*Breathe in deeply and fill your lungs from the
bottom to the top as if you are filling up a
water bottle.*
*As you breathe out, empty your lungs from the
top to the bottom.*
Let your breath be slow and steady.
*Observe the movements of your abdomen and
chest.*
*Stay connected to the wonderful release of tension
and the deep feeling of relaxation.*
*Let this feeling extend to every part of your
body.*

Let every breath be a vehicle to discover new ways to deepen your relaxation.
Pause.
Now redirect your full attention to your breath.
Bring all of your attention to the movement of your abdomen and chest as you breathe in and out.
Create no struggle with your breathing. Use the breath to release tensions.
Let the flow of breath be steady and uniform from bottom to top as much as possible.
Pause.
With each breath out, release any tension held in your body.
Let all the thoughts in your mind simply drift away.
Feel how every breath fills every cell and nerve with pulsating energy.
Now breathe normally and be still.
Pause.
Bring your total attention to the energy field in the form of sensation in your body.
Let all tension simply melt, drain away and dissolve into the fabric of the universe itself.
Bring all your attention to your eyebrow center.
Empty your body and mind and enter deeper levels of stillness and silence.
Pause.

INTEGRATE

Do absolutely nothing from now on.
Settle into the silent source of your being.
Pause.

Heaviness

We will now use our attention to communicate to the deepest parts of our body through intention alone. As I name a body part, feel the heaviness there sinking like a stone in the ocean.
Both feet...heavy...like stones.
Calves and knees...heavy...sinking deeper.
Thighs and hips...very heavy...like lead.
Abdomen, chest and back...gravity pulling them down to the source of everything.
Deeper still.
Shoulders, arms and palms...very...very...heavy.
Feel your entire head, heavy like a stone.
Give your body completely and totally to the omnipresent field of gravity.
Know that force is your home.
It binds every part of your being and the universe together.
Pause.
Now experience your whole body heavy like a rock.
Feel your whole body sinking deeper...and deeper.
The depths of your consciousness are as vast as the breadth of the universe.
Totally let go into the pull of gravity.

Sinking deeper into stillness and silent awareness.

LIGHTNESS

Now shift your attention, and as I name the part of your body, let all the heaviness drain away.
Let your body be buoyant and light, navigating the infinite space of consciousness like a fluffy cloud.
Both feet...limp and light.
Calves and knees...empty and free. Feel it.
Thighs and hips...hollow and weightless.
Abdomen, chest and back...light and empty.
Shoulders and arms and palms...floating.
Head...hollow...empty. Weightless.
Feel your whole body...empty...light and hollow.
Sense the emptiness of your body and silence of your mind.
Feel this moment.
Pause.

INTENTION

Know you are in a place of perfect balance in the stillness of consciousness.
Here your intentions and affirmations are actualized with effortless ease.

*Here is the space where you may let go of self-
 defeating patterns and habits that hold you
 back.*
*Here is the place where you can communicate
 directly with sickness and health.*
*It is where the metaphysical becomes physical
 and is the effortless root of your best self.*
Make your intention now.
Repeat it silently three times.
Pause.
*Allow it to go to the deepest levels of recognition
 with no hesitation.*
*Know that the truest form of your self honors and
 accepts your intention.*
*Know that your intention is part of the fabric of
 the universe.*
*Have faith and trust that it has been heard and is
 being acted upon by the source of all
 consciousness.*
There is no need for you to do anything about it.

Affirmations

*Allow your entire self to respond spontaneously
 and effortlessly to what I say.*
*I am at peace with myself as I am, and the world
 as it is.*
Pause.
*I replace resentment and regret with total accep-
 tance and forgiveness.*
Pause.

I return to the innate wisdom of my body to heal itself. I remain in restful awareness.
Pause.

Higher Self

Establish yourself in trust and faith to receive the wisdom, grace and protection from the depths of your being and the root of the universe itself.

The more often you go to the source, the easier it will be to return, and the longer you can stay there.

Feel the presence of the perfect wisdom that guides you.

Know that you are the universe aware of itself.

Feel the presence of the love of the source of everything and accept its blessings.

Embody it and spread it wherever you go.

Now you have prepared the base, you can carry out all of the interactions in life by drawing on the power within yourself and all that is outside.

It is all one.

If you have an area that you feel needs healing—physical, mental or emotional—allow this light and love to flow into that area now.

Pause.

Externalize

> *Now, begin to become aware of the rising and falling of the breath.*
> *Pause.*
> *Slowly...feel yourself begin to rise to the surface of awareness.*
> *Pause.*
> *Sense the body resting on the floor and the quality of the air as it touches the skin.*
> *Pause.*
> *Gradually, you can move as if you are waking from a restful sleep.*
> *Bend your knees and pull them close to your chest, rocking sideways gently.*
> *Take your time; do not hurry.*
> *Now, just turn onto your side and curl into a fetal position. Feel the safety, comfort and protection of this pose.*
> *Bring your intention into your awareness again.*
> *Change nothing.*
> *Every time you find yourself in reaction, you are empowered to replace it with your intention.*
> *Now you can gradually move and begin to sit up with your eyes closed.*
> *Continue to stay deep in this inner experience.*
> *Regardless of what you consciously recognize has or has not changed, know something deep within has shifted to connect with your intention.*
> *Become aware of your body...and bring a deep sense of peace and contentment with you as you bring your attention back to the body.*

Notice:

How relaxed the body is
How soft the breath is
How silent the mind is
How quiet the heartbeat is
Be still...and grateful.
Know you can easily enter here again and again.
Now you may gradually open your eyes.

ACKNOWLEDGMENTS

This book would not have been possible without the generous support of the community at kickstarter who were with me when I envisioned writing a book about short-midday naps and stayed along for the ride after I realized that I was really writing a book about dreaming.

Thank you so much to Aleš, Ali K, Alison Schiffern, Amanda, Andrea, Andrew Mason, Andrew Rowley, Angel, Anne O'Nymous, Anthony Renna, Arjan van der Schoot, Ben Ellsworth, Bjorn Peer Lorenzo-van der Meer, Brad Gray, Brad Lincoln, Brady, Brian Bigglesbee, Brian Coan, Brian Wolfe, Bryan Reid, Carrie Alyson, Chris La Tray, Chris Petrella, Chris Walker, Christian Peck, Collin Kachel, D Musch, Damian Lauria, Daniel Kramp, Daniel Wright, Derek Kedziora, dneder, Dspot19, Elle, Eric, Eric Schreiber, Erlend Hunstad, Evan, Girish Gupta, Greg Lehrmann, Hanna Brandli, Hudson Lindenberger, Jacob, James French, Jared Pogson, jason miklian, Jason Padvorac, Jeff, Jeffrey Freund, Jeffrey Pedelty, Jennifer Lahl & Matthew Eppinette, John, John Ferrari, John Odden, Jonathan Rubin, Jordan Barnes, Justin, Karen Capuzzi, karl, Larry Yu, Laura Brooks, Leland Krantz, Linda Carney, Lindsay Maluokapu'uwai, Luke Schmidt, M&M, Margaret, Marty McCabe, Mary McBride, Matt Bush, Matt McClain, Metin, Michael Chwalek, Mike Jarema, Mike Kranis, Neil, Oskar Lechner, Paul Stoneking, Pavel Krivenko, Peter Warren, Rebecca Hall, Rosylyn Rhee, Ryan Boykin, Ryan Findley, Ryan H,

Samantha Siberini, Sarah Boman, Scott Keneally, Scott Kenemore, Scott Whittlestone, Scotti Lechuga, Sebastien Zappa, Seth Fangboner, Sharon Desilets, Sherry, The Creative Fund by BackerKit, Thijs Perenboom, Tony Meyer, Travis Grimm, Troels Wittendorff Gram, Victor Frandsen, Vyga, Wendy Cook, William Ellison, William Lahaie.

REFERENCES

Benoit, Roland. "Our memory is not made for the past, but for the future." Max Planck Institute, 2016.

Call, Annie Payson. Power Through Repose. Project Gutenberg. Original publication 1891.

Clune, Michael. "Night Shifts," Harper's, 2022.

Cordi MJ, Schlarb AA, Rasch B. Deepening sleep by hypnotic suggestion. Sleep. 2014 Jun 1;37(6):1143-52, 1152A-1152F. doi: 10.5665/sleep.3778. PMID: 24882909; PMCID: PMC4015388.

Coren, Stanley. Sleep Thieves. The Free Press, 1993.

Desai, Kamini. Yoga Nidra. Lotus Press, 2017.

A. Roger Ekrich, At Day's Close: Night in Times Past: A History of Nighttime. W.W. Norton, 2005.

Ekirch AR. Segmented Sleep in Preindustrial Societies. Sleep. 2016 Mar 1;39(3):715-6. doi: 10.5665/sleep.5558. PMID: 26888454; PMCID: PMC4763365.

Gefter, Amanda. "What Are Dreams For?" The New Yorker, August 31, 2023.

Gladwell, Malcolm. Outliers: The Story of Success. Hachette, 2008.

Haar Horowitz, Adam. Incubating Dreams: Awakening Creativity. Massachusetts Institute of Technology. Master's Thesis, 2019.

Haar Horowitz, Adam. Dormio: Interfacing With Dreams. Massachusetts Institute of Technology. Ph.D. Dissertation, 2022.

Hersey, Tricia. Rest Is Resistance: A Manifesto. Little Brown, 2022.

Jacobson, Edmund. You Must Relax. McGraw-Hill, 1957.

Kjaer TW, Bertelsen C, Piccini P, Brooks D, Alving J, Lou HC. Increased dopamine tone during meditation-induced change of consciousness. Brain Res Cogn Brain Res. 2002 Apr;13(2):255-9. doi: 10.1016/s0926-6410(01)00106-9. PMID: 11958969.

Lou HC, Kjaer TW, Friberg L, Wildschiodtz G, Holm S, Nowak M. A 15O-H2O PET study of meditation and the resting state of normal consciousness. Hum Brain Mapp. 1999;7(2):98-105. doi: 10.1002/(SICI)1097-0193(1999)7:2<98::AID-HBM3>3.0.CO;2-M. PMID: 9950067; PMCID: PMC6873339.

Mednick, Sara. Take a Nap! Change Your Life. Random House, 2006.

Panda, Satchin. The Circadian Code. Rodale, 2018.

Pandi-Perumal SR, Spence DW, Srivastava N, Kanchibhotla D, Kumar K, Sharma GS, Gupta R, Batmanabane G. The Origin and Clinical Relevance of Yoga Nidra. Sleep Vigil. 2022;6(1):61-84. doi: 10.1007/s41782-022-00202-7. Epub 2022 Apr 23. PMID: 35496325; PMCID: PMC9033521.

Samson DR, Crittenden AN, Mabulla IA, Mabulla AZ, Nunn CL. Hadza sleep biology: Evidence for flexible sleep-wake patterns in hunter-gatherers. Am J Phys Anthropol. 2017 Mar;162(3):573-582. doi: 10.1002/ajpa.23160. Epub 2017 Jan 7. PMID: 28063234.

Samson, DR. The Human Sleep Paradox: The Unexpected Sleeping Habits of Homo sapiens. Annual Review of Anthropology, Vol. 50:259-274. https://doi.org/10.1146/annurev-anthro-010220-075523

Scheper-Hughes, Nancy. Death Without Weeping: The Violence of Everyday Life in Brazil.. University of California Press, 1993.

Strand, Clark. Waking up to the Dark:The Black Madonna's Gospel for An Age of Extinction and Collapse. Monkfish Book Publishing, 2002.

Tully J. Management of ADHD in Prisoners—Evidence Gaps and Reasons for Caution. Frontiers in Psychiatry. 2022 Mar 18;13:771525. doi: 10.3389/fpsyt.2022.771525. PMID: 35370827; PMCID: PMC8973692. https://www.ncbi.nlm.nih.gov/pmc/articles/PMC8973692/

Walker, Matthew. Why We Sleep: Unlocking the Power of Sleep and Dreams. Penguin Books, 2018.

Weil, Andrew. "What No One Wants to Know About Marijuana," The Natural Mind, 1986.

Wiseman, Richard. Night School: Wake Up to the Power of Sleep. McMillan, 2015.

www.ingramcontent.com/pod-product-compliance
Lightning Source LLC
Chambersburg PA
CBHW020029040426
42333CB00039B/727